MATH PREP for the
CUNY Elementary Algebra Final Exam
Workbook

Rachel Sturm-Beiss

Kingsborough Community College

Joshua Yarmish

Pace University

Kendall Hunt

p u b l i s h i n g   c o m p a n y

Cover image © Shutterstock, Inc. Used under license.

www.kendallhunt.com
Send all inquiries to:
4050 Westmark Drive
Dubuque, IA 52004-1840

Preface

This workbook is intended to prepare students of City University of New York for the "CUNY Elementary Algebra Final Exam" (CEAFE). Each chapter's sections have explanations, worked out examples, exercises, and worksheets intended for classroom use. Answers are provided for all exercises and worksheets. There is an appendix with three practice final exams.

The accompanying web-site, *cuny.mathbreeze.com*, includes automatically graded problem sets for each topic, coupled with videos that offer explanations and demonstrations meant assist students with the problem sets. Instructors can assign problem sets for home-work and view their students' progress online.

Rachel Sturm-Beiss
Joshua Yarmish

CHAPTER 1 : Integers, Exponents and Order of Operations

1A – Adding Integers

The **integers** are as follows: $\dots, -3, -2, -1, 0, 1, 2, 3, \dots$

The integers on the number line:

EXAMPLE 1: Adding on the number line.

A) $2 + 3 =$ **B)** $2 + (-3) =$ **C)** $-2 + (-3) =$ **D)** $-3 + 2 =$ **E)** $3 + (-2) =$

SOLUTION:

A)

$$2 + 3 = 5$$

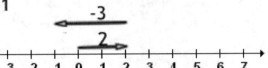

B)

$$2 + (-3) = -1$$

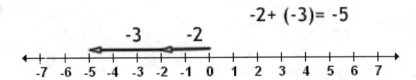

C)

$$-2 + (-3) = -5$$

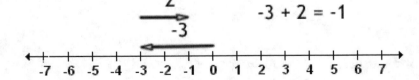

D)

$$-3 + 2 = -1$$

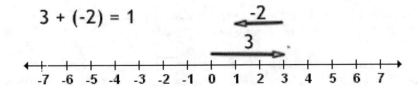

E)

$$3 + (-2) = 1$$

■

Definition: The **absolute value** of a non-negative number is the same as the number. The absolute value of a negative number is the number without its sign.

The absolute value of -2 is written as $|-2|$ and is equal to 2.

EXAMPLE 2: Evaluate the expression.

A) $|5|$ **B)** $|-6|$ **C)** $-|5|$ **D)** $-|-3|$

SOLUTION:

A) $|5| = 5$ **B)** $|-6| = 6$
C) $-|5| = -5$ **C)** $-|-3| = -3$ ∎

Rules for adding integers and signed numbers:

- **Adding two numbers with the same signs:**
  - Add their absolute values.
  - Use their common sign for the sign of the sum
- **Adding two numbers with different signs:**
  - Subtract the larger absolute value minus the smaller absolute value.
  - Use the sign of the number with the larger absolute value for the sum.

EXAMPLE 3: Add

A) $7 + 10$ **B)** $-30 + (-50)$ **C)** $7 + (-5)$

D) $-4 + 10$ **E)** $10 + (-12)$ **F)** $-20 + 15$

SOLUTION:

A) $7 + 10 = \mathbf{17}$

B) $-30 + (-50)$ Add their absolute values: $30 + 50 = 80$
                                  Use their common sign for the sum: $-30 + (-50) = \mathbf{-80}$

C) $7 + (-5)$ Subtract the larger(in absolute value) minus the smaller: $7 - 5 = 2$
                                  Use the sign of the number with larger absolute value: $7 + (-5) = \mathbf{2}$

D) $-4 + 10$ Subtract the larger (in absolute value) minus the smaller: $10 - 4 = 6$
                                  Use the sign of the number with larger absolute value: $-4 + 10 = \mathbf{6}$

E) $10 + (-12)$ Subtract the larger(in absolute value) minus the smaller: $12 - 10 = 2$

Use the sign of the number with larger absolute value: $10 + (-12) = -2$

F) $\quad -20 + 15 \qquad$ Subtract the larger absolute value minus the smaller: $20 - 15 = 5$
Use the sign of the number with larger absolute value: $-20 + 15 = -5$ ∎

1A – EXERCISES

For $1 - 6$, add on the number line.

1. $\quad 2 + 4$

2. $\quad 2 + (-4)$

3. $\quad -4 + 2$

4. $\quad -3 + (-4)$

5. $\quad -3 + 7$

6. $\quad 4 + (-7)$

For $7 - 24$, add.

7. $\qquad 15 + 25$

8. $\qquad -15 + (-5)$

9. $\qquad -30 + (-20)$

10. $\qquad -45 + (-15)$

11. $\qquad -55 + (-5)$

12. $\qquad -32 + (-12)$

13. $\qquad 15 + (-5)$

14. $\qquad 25 + (-12)$

15. $\qquad 35 + (-15)$

16. $\qquad 12 + (-14)$

17. $\qquad 25 + (-30)$

18. $\qquad 24 + (-28)$

19. $\qquad -30 + 50$

20. $\qquad -25 + 75$

21. $\qquad -12 + 14$

22. $\qquad -15 + 5$

23. $\qquad -25 + 10$

24. $\qquad -42 + 15$

1A – WORKSHEET: Adding Integers

For 1 – 5, add on the number line.

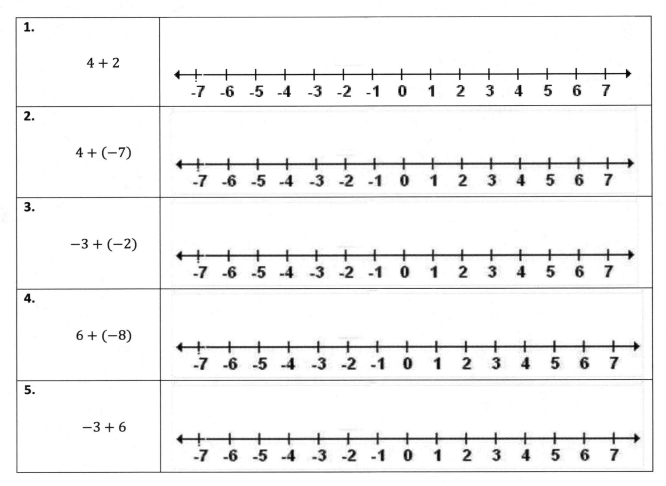

1. $4 + 2$	
2. $4 + (-7)$	
3. $-3 + (-2)$	
4. $6 + (-8)$	
5. $-3 + 6$	

For 6 - 23 , add.

6. $25 + 72$		**7.** $-32 + (-23)$		**8.** $-15 + (-54)$	
9. $-36 + (-35)$		**10.** $-73 + (-34)$		**11.** $-65 + (-32)$	

12. $\quad$ $70 + (-20)$	**13.** $\quad$ $15 + (-35)$	**14.** $\quad$ $60 + (-45)$
15. $\quad$ $43 + (-63)$	**16.** $\quad$ $34 + (-12)$	**17.** $\quad$ $45 + (-65)$
18. $\quad$ $-20 + 15$	**19.** $\quad$ $-15 + 40$	**20.** $\quad$ $-150 + 40$
21. $\quad$ $-80 + 100$	**22.** $\quad$ $-34 + 18$	**23.** $\quad$ $-45 + 90$

Answers:

1. 6	**2.** -3	**3.** -5	**4.** -2
5. 3	**6.** 97	**7.** -55	**8.** -69
9. -71	**10.** -107	**11.** -97	**12.** 50
13. -20	**14.** 15	**15.** -20	**16.** 22
17. -20	**18.** -5	**19.** 25	**20.** -110
21. 20	**22.** -16	**23.** 45	

1B – Subtracting Integers

Rule for subtracting integers: $a - b = a + (-b)$

EXAMPLE 1: Subtract

A) $5 - 10$ B) $-7 - 30$ **C)** $10 - (-5)$ **D)** $-15 - (-5)$

SOLUTION:

A) $5 - 10 = 5 + (-10) = -5$

B) $-7 - 30 = -7 + (-30) = -37$

C) $10 - (-5) = 10 + \left(-(-5)\right) = 10 + 5 = 15$
Notice that *when we subtract a negative we add a positive*.

D) $-15 - (-5) = -15 + 5 = -10$
Notice again that subtracting a negative is the same as adding a positive. ∎

1B – EXERCISES

1. $25 - 15$	**2.** $12 - 25$	**3.** $17 - 28$
4. $15 - 30$	**5.** $5 - 23$	**6.** $8 - 17$
7. $3 - (-20)$	**8.** $12 - (-13)$	**9.** $14 - (-25)$
10. $28 - (-13)$	**11.** $42 - (-26)$	**12.** $33 - (-14)$
13. $25 - (-16)$	**14.** $13 - (-32)$	**15.** $15 - (-45)$
16. $-20 - 15$	**17.** $-45 - 32$	**18.** $-32 - 24$
19. $-18 - 17$	**20.** $-23 - 15$	**21.** $-41 - 28$
22. $-33 - 25$	**23.** $-48 - 27$	**24.** $-36 - 48$
25. $-12 - (-15)$	**26.** $-25 - (-32)$	**27.** $-15 - (-45)$
28. $-32 - (-75)$	**29.** $-43 - (-12)$	**30.** $-50 - (-15)$
31. $5 - 7 + 10 - 8$	**32.** $-8 - 10 + 12 - 6$	**33.** $3 - 15 + 20 - 5$

1B – WORKSHEET: Subtracting Integers

1. $30 - 15$	**2.** $13 - 25$	**3.** $15 - 28$
4. $20 - 30$	**5.** $7 - 23$	**6.** $5 - 17$
7. $10 - (-20)$	**8.** $15 - (-13)$	**9.** $12 - (-25)$
10. $25 - (-14)$	**11.** $43 - (-27)$	**12.** $35 - (-16)$
13. $30 - (-16)$	**14.** $17 - (-33)$	**15.** $20 - (-45)$
16. $-25 - 15$	**17.** $-47 - 32$	**18.** $-35 - 23$
19. $-19 - 14$	**20.** $-27 - 15$	**21.** $-45 - 26$
22. $-35 - 25$	**23.** $-58 - 27$	**24.** $-56 - 48$
25. $-12 - (-25)$	**26.** $-25 - (-42)$	**27.** $-15 - (-43)$
28. $-32 - (-15)$	**29.** $-43 - (-13)$	**30.** $-40 - (-15)$
31. $10 - 15 + 12 - 20$	**32.** $-8 - 5 - 10 + 4$	**33.** $-7 - 12 + 10 - 13$

Answers:

1. 15	**2.** -12	**3.** -13	**4.** -10	**5.** -16
6. -12	**7.** 30	**8.** 28	**9.** 37	**10.** 39
11. 70	**12.** 51	**13.** 46	**14.** 50	**15.** 65
16. -40	**17.** -79	**18.** -58	**19.** -33	**20.** -42
21. -71	**22.** -60	**23.** -85	**24.** -104	**25.** 13
26. 17	**27.** 28	**28.** -17	**29.** -30	**30.** -25
31. -13	**32.** -19	**33.** -22		

1C – Multiplying and Dividing Integers

Rules for multiplying integers and signed numbers:

- The product of two numbers with the same sign is a positive number.
- The product of two numbers with different signs is a negative number.

EXAMPLE 1:

A) $(-5)(12)$ **B)** $(7)(-6)$ **C)** $(-11)(-5)$ **D)** $(12)(6)$ **E)** $(3)(-2)(5)(-4)$

SOLUTION:

A) $(-5)(12) = -60$ **B)** $(7)(-6) = -42$

C) $(-11)(-5) = 55$ **D)** $(12)(6) = 72$ **E)** $(3)(-2)(5)(-4) = (-6)(5)(-4) = (-30)(-4) = 120$ ∎

Rules for dividing integers and signed numbers:

- The quotient of two numbers with the same sign is a positive number.
- The quotient of two numbers with different signs is a negative number.

EXAMPLE 2:

A) $\dfrac{25}{5}$ **B)** $\dfrac{-30}{6}$ **C)** $\dfrac{60}{-10}$ **D)** $\dfrac{-75}{-25}$

SOLUTION:

A) $\dfrac{25}{5} = 5$ **B)** $\dfrac{-30}{6} = -5$ **C)** $\dfrac{60}{-10} = -6$ **D)** $\dfrac{-75}{-25} = 3$ ∎

1C – EXERCISES:

1. $(-12)(7)$ **2.** $(15)(-4)$ **3.** $(30)(-7)$ **4.** $(14)(-6)$

5. $(32)(-5)$ **6.** $(7)(-15)$ **7.** $(-20)(-5)$ **8.** $(-45)(-23)$

9. $(-24)(-35)$ **10.** $(-52)(-34)$ **11.** $(46)(-32)$ **12.** $(23)(-76)$

13. $\dfrac{-25}{5}$ **14.** $\dfrac{36}{-18}$ **15.** $\dfrac{-75}{15}$ **16.** $\dfrac{60}{-12}$

17. $\dfrac{-10}{5}$ **18.** $\dfrac{75}{-25}$ **19.** $\dfrac{42}{-6}$ **20.** $\dfrac{-56}{7}$

21. $\frac{-45}{-9}$ **22.** $\frac{-36}{-12}$ **23.** $\frac{-40}{-8}$ **24.** $\frac{-16}{-4}$

25. $\frac{-22}{-11}$ **26.** $\frac{-55}{-5}$ **27.** $\frac{-72}{-12}$ **28.** $\frac{-80}{-16}$

29. $(-5)(3)(-2)$ **30.** $(-4)(-3)(5)$ **31.** $(-7)(2)(-4)(-3)$

1C – WORKSHEET: Multiplying and Dividing Integers

1. $(-14)(8)$	**2.** $(18)(-5)$	**3.** $(37)(-8)$	**4.** $(24)(-6)$
5. $(42)(-7)$	**6.** $(8)(-25)$	**7.** $(-23)(-6)$	**8.** $(-55)(-33)$
9. $(-25)(-45)$	**10.** $(-62)(-32)$	**11.** $(56)(-33)$	**12.** $(43)(-75)$
13. $\frac{-35}{5}$	**14.** $\frac{54}{-18}$	**15.** $\frac{-45}{15}$	**16.** $\frac{60}{-4}$
17. $\frac{35}{-5}$	**18.** $\frac{125}{-25}$	**19.** $\frac{48}{-6}$	**20.** $\frac{-63}{7}$
21. $\frac{-54}{-9}$	**22.** $\frac{-48}{-12}$	**23.** $\frac{-56}{-8}$	**24.** $\frac{-64}{-4}$
25. $\frac{-33}{-11}$	**26.** $\frac{-77}{-7}$	**27.** $\frac{-84}{-12}$	**28.** $\frac{-96}{-16}$
29. $(5)(-2)(-3)$	**30.** $(-7)(2)(5)$	**31.** $(-3)(2)(-4)(5)$	**32.** $(-7)(-2)(-3)(5)$

Answers:

1. -112	**2.** -90	**3.** -296	**4.** -144	**5.** -294
6. -200	**7.** 138	**8.** 1815	**9.** 1125	**10.** 1984
11. -1848	**12.** -3225	**13.** -7	**14.** -3	**15.** -3
16. -15	**17.** -7	**18.** -5	**19.** -8	**20.** -9
21. 6	**22.** 4	**23.** 7	**24.** 16	**25.** 3
26. 11	**27.** 7	**28.** 6	**29.** 30	**30.** -70
31. 120	**32.** -210			

1D – Exponents and Order of Operations

The repeated multiplication, $2 \cdot 2 \cdot 2 \cdot 2 \cdot 2$, can be expressed as:

$$2 \cdot 2 \cdot 2 \cdot 2 \cdot 2 = 2^5$$

In the expression 2^5, 2 is the **base**, and 5 is the **exponent**.

The expression 2^5 is read: "2 *to the* 5th *power*" or "2 *to the* 5th". **Power** is another word for exponent.

The expression 2^0 is equal to 1. In general, $\boldsymbol{a^0 = 1}$, for all $a \neq 0$. (0^0 is undefined.)

EXAMPLE 1: Write using exponential notation.

A) $3 \cdot 3 \cdot 3 \cdot 3$ **B)** $5 \cdot 5 \cdot 5 \cdot 7 \cdot 7 \cdot 7 \cdot 7$ **C)** $(-2)(-2)(-2)(-2)$ **D)** 4

SOLUTION:

A) $3 \cdot 3 \cdot 3 \cdot 3 = 3^4$ **B)** $5 \cdot 5 \cdot 5 \cdot 7 \cdot 7 \cdot 7 \cdot 7 = 5^3 7^4$

C) $(-2)(-2)(-2)(-2) = (-2)^4$ **D)** 4^1 ■

EXAMPLE 2: Write in expanded form.

A) 4^3 **B)** $2 \cdot 3^3$ **C)** $5^2 7^3$ **D)** $(-3)^3$

SOLUTION:

A) $4^3 = 4 \cdot 4 \cdot 4$ **B)** $2 \cdot 3^3 = 2 \cdot 3 \cdot 3 \cdot 3$ **C)** $5^2 7^3 = 5 \cdot 5 \cdot 7 \cdot 7 \cdot 7$ **D)** $(-3)^3 = (-3)(-3)(-3)$

                                                                     ■

EXAMPLE 3: Evaluate.

A) 2^3 **B)** $3 \cdot 2^2$ **C)** $(-2)^4$ **D)** $(-3)^3$ **E)** 5^0

SOLUTION:

A) 8   **B)** $3 \cdot 4 = 12$ **C)** 16   **D)** -27 **E)** 1 ■

If an expression contains more than one operation, we do the operations in a specific order.

The Order of Operations:

- Evaluate expressions in parenthesis first.
- Evaluate expressions with exponents.
- Multiply or divide from left to right.
- Add or subtract from left to right.

EXAMPLE 4: Evaluate the mathematical expression.

1. $5 + 3 \times 2$
2. $2 \cdot 3^2$
3. $8 - 2(3)$
4. $6 - 2(5 - 3)$
5. $2 - 5(4 - 8) \div 2$
6. $(-1)3^2$
7. -3^2
8. $3(-2)^4$
9. $\frac{10+4}{2}$
10. $-2[3 + 2(5 - 8)] + 10$
11. $10 - (5 - 2)^2$
12. $3(2 - 5)^2 + 2(-5)$
13. $2 - |3 - 7|$
14. $-5|5 - 8| + 3$

SOLUTION:

1. $5 + 3 \times 2 = 5 + 6 = 11$

2. $2 \cdot 3^2 = 2 \cdot 9 = 18$

3. $8 - 2(3) = 8 - 6 = 2$

4. $6 - 2(5 - 3) = 6 - 2(2) = 6 - 4 = 2$

5. $2 - 5(4 - 8) \div 2 = 2 - 5(-4) \div 2 = 2 + 20 \div 2 = 2 + 10 = 12$

6. $(-1)3^2 = (-1) \cdot 9 = -9$

7. $-3^2 = -9$ We can expand: $-3^2 = -3 \cdot 3 = -9$. Or, $-3^2 = (-1)3^2 = (-1)9 = -9$.

8. $3(-2)^4 = 3(16) = 48$

9. $\frac{10+4}{2} = \frac{14}{2} = 7$ Notice that the numerator (and denominator) groups an expression in the same way as parenthesis.

10. $-2[3 + 2(5 - 8)] + 10 = -2[3 + 2(-3)] + 10 = -2[3 - 6] + 10 = -2[-3] + 10 = 6 + 10 = 16$

11. $10 - (5 - 2)^2 = 10 - (3)^2 = 10 - 9 = 1$

12. $3(2 - 5)^2 + 2(-5) = 3(-3)^2 + 2(-5) = 3(9) + 2(-5) = 27 - 10 = 17$

13. $2 - |3 - 7| = 2 - |-4| = 2 - 4 = -2$ **14.** $-5|5 - 8| + 3 = -5|-3| + 3 = -5(3) + 3 = -12$ ■

1D – EXERCISES

For 1 – 8, write using exponential notation.

1. $4 \cdot 4 \cdot 4$ **2.** $3 \cdot 5 \cdot 5 \cdot 5 \cdot 5$ **3.** $2 \cdot 2 \cdot 2 \cdot 7 \cdot 7$ **4.** $(-4)(-4)(-4)$

5. $-5 \cdot 5 \cdot 5 \cdot 5$ **6.** $(-5)(-5)(-5)(-5)$ **7.** $-2 \cdot 3 \cdot 3$ **8.** $-7 \cdot 7 \cdot 6 \cdot 6 \cdot 6$

For 9 - 16 , write in expanded form.

9. 3^5 **10.** $2 \cdot 5^3$ **11.** $2^3 7^4$ **12.** $-3 \cdot 5^2$

13. -3^5 **14.** $(-3)^5$ **15.** $-2^3 5^4$ **16.** $(-7)^3$

For 17 - 48 , evaluate.

17. $3 + 7 - 5$ **18.** $4 - 5 - 7 + 2$

19. $2 + 5 \times 3$ **20.** $(2 + 5) \times 3$

21. $\frac{10 + 2}{4 + 2}$ **22.** $10 - 2(3)$

23. $3 - 7(-4)$ **24.** $2 - 5(3 - 10)$

25. $6 + 5(2 - 4)$ **26.** $4 - 2(3 - 7) + 5$

27. $6 - 3(9 - 7)$ **28.** $3 + 4(5 - 8) - 6$

29. $-2[6 + 5(2 - 5)]$ **30.** 5^3

31. -5^2 **32.** $(-2)3^2$

33. $(-5)^2$ **34.** $(-2)^3$

35. -3^3 **36.** $(-3)^3$

37. $-3 \cdot 5^2$ **38.** -2^4

39. $3 + (5 - 2)^2$ **40.** $2 + 3(4 - 2)^3$

41. $5 + 2(3 - 5)^2$ **42.** $3^0 + 2(5 - 10)^2$

43. $3 - 2(3 - 5)^2$ **44.** $2 - 5(2 - 4)^3$

45. $15 - (5 - 3)^2 + 2(-3)$ **46.** $10 + (4 - 2)^3 + 3(-2)$

47. $3(-5) + (2 - 5)^3 + 3(-4)$ **48.** $15 - (2 - 7)^2 + 3(-5) + 3^0$

49. $-3|5 - 10|$

50. $3 - 4|2 - 5|$

51. $5 + 2|3 - 7|$

52. $10 - 2|4 - 8|$

1D – WORKSHEET: Exponents and Order of Operations

For 1 – 8, write using exponential notation.

1. $3 \cdot 3 \cdot 3$	**2.** $2 \cdot 7 \cdot 7 \cdot 7$	**3.** $4 \cdot 8 \cdot 8 \cdot 7 \cdot 7$	**4.** $(-3)(-3)(-3)$
5. $-7 \cdot 7 \cdot 7$	**6.** $(-4)(-4)(-4)$	**7.** $-5 \cdot 3 \cdot 3$	**8.** $-7 \cdot 7 \cdot 8 \cdot 8 \cdot 8$

For 9 - 16 , write in expanded form.

9. 7^5	**10.** $2 \cdot 7^3$	**11.** $4^3 8^4$	**12.** -9^2
13. -4^5	**14.** $(-4)^5$	**15.** $-2^3 8^4$	**16.** $(-6)^3$

For 17 - 48 , evaluate.

17. $4 + 10 - 5$	**18.** $8 - 2 - 7 + 2$
19. $2 + 7 \times 3$	**20.** $(2 + 7) \times 3$
21. $\dfrac{12+2}{4+3}$	**22.** $15 - 2(3)$
23. $3 - 5(-4)$	**24.** $8 - 5(2 - 10)$
25. $7 + 3(2 - 4)$	**26.** $9 - 2(4 - 7) + 6$

27. $8 - 2(9 - 7)$	**28.** $5 + 2(5 - 8) - 7$
29. $-2[7 + 4(3 - 5)]$	**30.** 4^3
31. -4^2	**32.** $(-2)4^2$
33. $(-4)^2$	**34.** $(-3)^3$
35. -4^3	**36.** $(-4)^3$
37. $-2 \cdot 5^2$	**38.** -3^4
39. $3 + (7 - 2)^2$	**40.** $4 + 3(5 - 2)^3$
41. $10 + 2(3 - 5)^2$	**42.** $4 + 2(6 - 10)^2$
43. $2^0 - 4(3 - 5)^2$	**44.** $5 - 3(2 - 4)^3$
45. $10 - (5 - 3)^2 + 2(-4)$	**46.** $15 + (4 - 2)^3 - 3(-2)$
47. $3(-4) + (2 - 5)^3 + 5(-4)$	**48.** $12 - (3 - 7)^2 + 3(-4) + 2^0$

49. $3 - 2|7 - 5|$	**50.** $-5|7 - 10| + 4$

Answers:

1.	3^3	**2.**	$2 \cdot 7^3$	**3.**	$4 \cdot 8^2 \cdot 7^2$
4.	$(-3)^3$	**5.**	-7^3	**6.**	$(-4)^3$
7.	$-5 \cdot 3^2$	**8.**	$-7^2 \cdot 8^3$	**9.**	$7 \cdot 7 \cdot 7 \cdot 7 \cdot 7$
10.	$2 \cdot 7 \cdot 7 \cdot 7$	**11.**	$4 \cdot 4 \cdot 4 \cdot 8 \cdot 8 \cdot 8 \cdot 8$	**12.**	$-9 \cdot 9$
13.	$-4 \cdot 4 \cdot 4 \cdot 4 \cdot 4$	**14.**	$(-4) \cdot (-4) \cdot (-4) \cdot (-4) \cdot (-4)$	**15.**	$-2 \cdot 2 \cdot 2 \cdot 8 \cdot 8 \cdot 8 \cdot 8$
16.	$(-6) \cdot (-6) \cdot (-6)$	**17.**	9	**18.**	1
19.	23	**20.**	27	**21.**	2
22.	9	**23.**	23	**24.**	48
25.	1	**26.**	21	**27.**	4
28.	-8	**29.**	2	**30.**	64
31.	-16	**32.**	-32	**33.**	16
34.	-27	**35.**	-64	**36.**	-64
37.	-50	**38.**	-81	**39.**	28
40.	85	**41.**	18	**42.**	36
43.	-15	**44.**	29	**45.**	-2
46.	29	**47.**	-59	**48.**	-15
49.	-1	**50.**	-11		

1 – Answers to Exercises

Section A

1.	6	**2.**	-2	**3.**	-2	**4.**	-7	**5.**	4
6.	-3	**7.**	40	**8.**	-20	**9.**	-50	**10.**	-60
11.	-60	**12.**	-44	**13.**	10	**14.**	13	**15.**	20
16.	-2	**17.**	-5	**18.**	-4	**19.**	20	**20.**	50
21.	2	**22.**	-10	**23.**	-15	**24.**	-27		

Section B

1.	10	**2.**	-13	**3.**	-11	**4.**	-15	**5.**	-18
6.	-9	**7.**	23	**8.**	25	**9.**	39	**10.**	41
11.	68	**12.**	47	**13.**	41	**14.**	45	**15.**	60
16.	-35	**17.**	-77	**18.**	-56	**19.**	-35	**20.**	-38
21.	-69	**22.**	-58	**23.**	-75	**24.**	-84	**25.**	3
26.	7	**27.**	30	**28.**	43	**29.**	-31	**30.**	-35
31.	0	**32.**	-12	**33.**	3					

Section C

1.	-84	**2.**	-60	**3.**	-210	**4.**	-84
5.	-160	**6.**	-105	**7.**	100	**8.**	1035
9.	840	**10.**	1768	**11.**	-1472	**12.**	-1748
13.	-5	**14.**	-2	**15.**	-5	**16.**	-5
17.	-2	**18.**	-3	**19.**	-7	**20.**	-8
21.	5	**22.**	3	**23.**	5	**24.**	4
25.	2	**26.**	11	**27.**	6	**28.**	5
29.	30	**30.**	60	**31.**	-168		

Section D

1.	4^3	**2.**	$3 \cdot 5^4$	**3.**	$2^3 \cdot 7^2$
4.	$(-4)^3$	**5.**	-5^4	**6.**	$(-5)^4$
7.	$-2 \cdot 3^2$	**8.**	$-7^2 \cdot 6^3$	**9.**	$3 \cdot 3 \cdot 3 \cdot 3 \cdot 3$
10.	$2 \cdot 5 \cdot 5 \cdot 5$	**11.**	$2 \cdot 2 \cdot 2 \cdot 7 \cdot 7 \cdot 7 \cdot 7$	**12.**	$-3 \cdot 5 \cdot 5$
13.	$-3 \cdot 3 \cdot 3 \cdot 3 \cdot 3$	**14.**	$(-3)(-3)(-3)(-3)(-3)$	**15.**	$-2 \cdot 2 \cdot 2 \cdot 5 \cdot 5 \cdot 5 \cdot 5$
16.	$(-7)(-7)(-7)$				

17.	5	**18.**	-6	**19.**	17	**20.**	21	**21.**	2	**22.**	4
23.	31	**24.**	37	**25.**	-4	**26.**	17	**27.**	0	**28.**	-15
29.	18	**30.**	125	**31.**	-25	**32.**	-18	**33.**	25	**34.**	-8
35.	-27	**36.**	-27	**37.**	-75	**38.**	-16	**39.**	12	**40.**	26
41.	13	**42.**	51	**43.**	-5	**44.**	42	**45.**	5	**46.**	12
47.	-54	**48.**	-24	**49.**	-15	**50.**	-9	**51.**	13	**52.**	2

CHAPTER 2: Introducing Algebraic Expressions and Equations

2A - Introducing Variables and Simple Algebraic Expressions

The New York City subway fare is $2.75. The fare for two people is 2($2.75). The cost of three rides and a new metro card (a new card costs $1.00) is 3($2.75) + $1.00.

Since the subway fare may change, let x represent the fare. Now, the fare for two people is represented by the expression $2x$, and the cost of three rides and a new metro card is represented by $3x + 1$.

EXAMPLE 1: Let x represent the subway fare. The cost of two rides if a new card is purchased is $2x + 1$. Find the cost if the fare is $4.50.

SOLUTION: $2(\$4.50) + 1 = \$9.00 + \$1.00 = \$10.00$ ∎

A letter that represents a number is called a **variable**. An expression containing variables and numbers related through algebraic operations $(+, -, \times, \div, and\ exponentiation)$ is called an **algebraic expression**. For example, $2x + 1$ is an algebraic expression. In this expression 2 is a **coefficient** of x, and 1 is a **constant coefficient**. The expression has two **terms**, $2x$ and 1.

We *evaluate* algebraic expressions by substituting values for the variables. For example, if $x = 3$, then $2x + 1 = 2(3) + 1 = 6 + 1 = 7$.

EXAMPLE 2: Evaluate the following expressions at the given values.

A) $3x + 7;\ x = 2$

B) $5x - 3; x = -4$

C) $2 - 7x; x = -3$

D) $3x^2 - 2x + 4;\ x = 4$

E) $7 - 3(5 - 2x);\ x = -2$

SOLUTION:

A) $3(2) + 7 = 6 + 7 = 13$

B) $5(-4) - 3 = -20 - 3 = -23$

C) $2 - 7(-3) = 2 + 21 = 23$

D) $3(4)^2 - 2(4) + 4 = 3(16) - 8 + 4 = 44$

E) $7 - 3\big(5 - 2(-2)\big) = 7 - 3(9) = 7 - 27 = -20$ ∎

EXAMPLE 3: Evaluate the expressions at the given variables.
  A) $xy + 2x^2y; x = 2, y = 3$ **B)** $a^2b - ab + 2b^2; a = -2, b = 4$
  C) $3x^2 - 2xy + 5y^2; x = -5, y = 2$

SOLUTION: **A)** $(2)(3) + 2(2)^2(3) = 6 + 2(4)(3) = 6 + 24 = 30$
  B) $(-2)^2(4) - (-2)(4) + 2(4)^2 = (4)(4) - (-2)(4) + 2(16) = 16 + 8 + 32 = 56$
  C) $3(-5)^2 - 2(-5)(2) + 5(2)^2 = 3(25) + 20 + 5(4) = 75 + 20 + 20 = 115$ ∎

Simplifying algebraic expressions

We simplify algebraic expressions by opening parenthesis and combining terms. For example, $3x + 2x$ is equal to $5x$, and $3(2 + 4x)$ is equal to $6 + 12x$.

Notice that we add the x-terms by adding the coefficients. This follows from the distributive law. For example,

$$3x + 2x = (3 + 2)x = 5x \quad \leftarrow \quad distributive\ law\ .$$

EXAMPLE 4: Simplify the following expressions.

A) $10x - 3x$

B) $5x + 3x + 10 + 12$

C) $3x - 10 - 2x + 5$

D) $3(5x + 10)$

E) $2(5 - 4x) + 7$

F) $-4(2 - 5x)$

G) $5 - 2(3 - 7x) + 10$

H) $12\left(\frac{x}{3} - \frac{3x}{4} + \frac{5}{6}\right)$

Solution:

A) $7x$

B) $8x + 22$

C) $x - 5$

D) $15x + 30$

E) $17 - 8x$

F) $-8 + 20x$

G) $9 + 14x$

H) $\dfrac{12}{1} \cdot \dfrac{x}{3} - \dfrac{12}{1} \cdot \dfrac{3x}{4} + \dfrac{12}{1} \cdot \dfrac{5}{6} = 4x - 9x + 10 =$
  $-5x + 10$ ∎

2A - EXERCISES

For 1 - 10, evaluate the algebraic expression at the given value of variable x.

1. $7x + 2, x = 5$

2. $3 - 5x, x = 2$

3. $3x + 4, x = -2$

4. $4 - 7x, x = -2$

5. $2x^2, x = -3$

6. $3x^2 - 5x - 2, x = -4$

7. $3 - 2(7 - x), x = -3$

8. $2x^2 - 7x - 2, x = 3$

9. $3 - 2(5 - 3x)$, $x = 4$ **10.** $4x^2 - 2x$, $x = -3$

For 11 – 20, simplify the algebraic expression.

11. $10 - 3x + 20 - 5x$ **12.** $3 - 2x - 7x + 4$

13. $5(3 - 4x) + 10$ **14.** $-2(4 - 3x) + 7$

15. $2 - 5(10 - 4x)$ **16.** $3 - 2(5 + 4x) - 7$

17. $7x - 5(4 - 4x) + 2$ **18.** $5x + 2(3 - 10x)$

19. $7x + 2(5 + 3x) - 4$ **20.** $3 - 4(8x - 5)$

For 21 – 26, evaluate the expression at the given values.

21. $3a - 2b$; $a = 2, b = 5$ **22.** $a^2 b + 2ab$; $a = -3, b = 2$

23. $3x - 2xy + y^2$; $x = -2, y = 5$ **24.** $10 - 2xy + 3x^2$; $x = 3, y = -4$

25. $2x^2 - 3y^2 + xy$; $x = 4, y = -2$ **26.** $3 - 4xy + 2x^2 y$; $x = -3, y = 2$

For 27 – 30, simplify.

27. $3\left(\frac{2x}{3} + \frac{1}{3}\right)$ **28.** $6\left(\frac{5x}{6} - \frac{x}{3}\right)$

29. $10\left(\frac{2x}{5} - \frac{1}{2}\right)$ **30.** $12\left(\frac{5x}{6} + \frac{x}{4} - \frac{2}{3}\right)$

2A – WORKSHEET: Introducing Variables and Simple Algebraic Expressions

For 1 – 10, evaluate the algebraic expression at the give value of the variable

1. $7 - 5x$, $x = 4$	**2.** $3 - 8x$, $x = -2$
3. $5x - 10$, $x = 7$	**4.** $3x^2 - 2x + 5$, $x = 4$
5. $2x - 3x^2$, $x = -3$	**6.** $3 - 5(2 - x)$, $x = 5$
7. $2 - 4(7 + 2x)$, $x = 2$	**8.** $3 - 2(4 - 5x)$, $x = -4$
9. $5x^2 - 10$, $x = -3$	**10.** $3 - 5x^2$, $x = 5$

For 11 – 20, simplify the algebraic expression

11. $10 + 7x - 30 - 5x$	**12.** $3x + 7x - 50 + 10$
13. $4(3x + 5)$	**14.** $-2(6 + 7x)$
15. $3(4 - 2x) + 10 - 5x$	**16.** $3 - 5(2 - 4x) + 3$
17. $7x - 4x - 2(3 - 5x)$	**18.** $2x - 5(8 - 3x)$
19. $-4(3 - 2x) - 3$	**20.** $3x - 2(5 - 6x)$

For 21 – 26, evaluate the expression at the given values.

21. $5a - 2b$; $a = 3, b = 2$	**22.** $ab^2 + 3ab$; $a = -2, b = 3$

23. $4x - 3xy + x^2$; $x = -3$, $y = 2$	**24.** $15 - 3xy + 2x^2$; $x = -3$, $y = 4$
25. $5x^2 - 2y^2 + xy$; $x = 3$, $y = -2$	**26.** $20 - 5xy + x^2y$; $x = -2$, $y = 3$

For $27 - 30$, simplify.

27. $5\left(\frac{3x}{5} + \frac{2}{5}\right)$	**28.** $6\left(\frac{7x}{6} - \frac{4x}{3}\right)$
29. $10\left(\frac{3x}{5} - \frac{7}{2}\right)$	**30.** $12\left(\frac{3x}{4} + \frac{x}{6} - \frac{5}{12}\right)$

Answers: 1. -13 **2.** 19 **3.** 25 **4.** 45 **5.** -33 **6.** 18 **7.** -42 **8.** -45 **9.** 35 **10.** -122
11. $-20 + 2x$ **12.** $10x - 40$ **13.** $12x + 20$ **14.** $-12 - 14x$ **15.** $22 - 11x$ **16.** $-4 + 20x$
17. $13x - 6$ **18.** $17x - 40$ **19.** $-15 + 8x$ **20.** $15x - 10$ **21.** 11 **22.** -36 **23.** 15 **24.** 69
25. 31 **26.** 62 **27.** $3x + 2$ **28.** $-x$ **29.** $6x - 35$ **30.** $11x - 5$

2B - Solving Linear Equations

Two algebraic expressions separated by an equal sign are called an **algebraic equation**. The following are examples of algebraic equations:

$$3x + 4 = 6x - 2$$

$$x^2 - 2 = 7$$

A **linear expression** (in one variable x) is one that can be simplified to the form: $ax + b$, where a and b are constants. A **linear equation** is two linear expressions separated by an equal sign. In the above examples, the first equation is linear and the second is not linear.

A **solution** of an equation is a value of the variable that makes both sides equal. For the above examples,

$3x + 4 = 6x - 2$ has solution $x = 2$, since $3(2) + 4 = 6(2) - 2$, and

$x^2 - 2 = 7$ has solution $x = 3$, since $(3)^2 - 2 = 7$ (note, $x = -3$ is another solution).

A solution of an equation is also called a **root**.

Two equations are **equivalent** if they have the same solution(s). The following basic operations yield equivalent equations.

Addition operation:	Adding the same expression to both sides of an equation.
Multiplication operation:	Multiplying (or dividing) both sides of an equation by the same non-zero expression.

We solve a linear equation by:

1. Simplifying both sides of the equation.

2. Use the addition and multiplication basic operations to isolate x on one side of the equation.

EXAMPLE 1: Solve for x

A) $3x = 12$

B) $2x + 5 = 15$

C) $3x - 2 = 5x + 10$

D) $3(5 - x) = 2x - 10$

E) $12 - 4x = -2(x - 5)$

F) $7 + 3x = -3(4 - 5x)$

G) $5(x - 3) = 3x + 2$

H) $\frac{2x}{5} + \frac{3}{10} = \frac{1}{5}$

SOLUTION:

A) $3x = 12$

$\dfrac{3x}{3} = \dfrac{12}{3}$ Divide both sides by 3.

$x = 4$

B) $2x + 5 = 15$

$\quad\quad -5 \quad -5$ Subtract 5 from both sides.

$\dfrac{2x}{2} = \dfrac{10}{2}$ Divide both sides by 2.

$x = 5$

C) $x = -6$ **D)** $x = 5$ **E)** $x = 1$ **F)** $x = \dfrac{19}{12}$ **G)** $x = \dfrac{17}{2}$

H) $\dfrac{10}{1}\left(\dfrac{2x}{5} + \dfrac{3}{10}\right) = \left(\dfrac{1}{5}\right)\dfrac{10}{1}$ Multiply both sides by 10 to clear out the denominators.

$4x + 3 = 2$

$4x = -1$

$x = -\dfrac{1}{4}$ ∎

If an equation has more than one variable we can solve for one of the variables. If the equation is linear in variable x, then we can solve for x the same way we solve a linear equation in one variable.

EXAMPLE 2: Solve for the indicated variable.

A) $z = 2x - 5y; \;\; solve\;for\;x$

B) $A = 3b + c; \;\; solve\;for\;b$

C) $V = \dfrac{1}{3}abh; \;\; solve\;for\;h$

D) $w = 3x + 4y; \;\; solve\;for\;y$

SOLUTION:

A) $2x = z + 5y, \quad x = \dfrac{z+5y}{2}$

B) $3b = A - c, \quad b = \dfrac{A-c}{3}$

C) $h = \dfrac{3V}{ab}$

D) $w - 3x = 4y, \quad y = \dfrac{w-3x}{4}$ ∎

2B - EXERCISES

For 1 - 14, solve the equation for the variable.

1. $5x = 10$ **2.** $x + 5 = 10$

3. $2x + 5 = 9$ **4.** $3x - 4 = 5$

5. $5n - 3 = 2n + 7$ **6.** $2 - 4n = 7n + 3$

7. $3(2x + 4) = 7x - 5$ **8.** $-2(3 - 4x) = 5x + 2$

9. $5 - 3(2 - 4n) = 10$ **10.** $3 - 4(2 - 5x) = -2(4 - x)$

11. $\frac{2}{3}x = 10$ **12.** $\frac{x}{5} = \frac{3}{10}$

13. $\frac{2x}{5} = \frac{4}{7}$ **14.** $\frac{3x}{2} = \frac{5}{3}$

For 15 - 22, solve for the indicated variable.

15. $2x + 3y = z;\ \ solve\ for\ y$ **16.** $ax + by = c;\ \ solve\ for\ y$

17. $3xy = z;\ \ solve\ for\ x$ **18.** $\frac{2}{3}xy = z;\ \ solve\ for\ x$

19. $A = \frac{1}{2}bh;\ \ solve\ for\ h$ **20.** $2x - 5y = 3z;\ \ solve\ for\ x$

21. $P = 2l + 2w;\ \ solve\ for\ w$ **22.** $\frac{2a}{b} = \frac{7c}{d}\ ;\ \ solve\ for\ c$

For $23 - 30$, first simplify by expressing without fractions, then solve for x.

23. $\frac{2x}{3} + \frac{1}{3} = \frac{2}{3}$ **24.** $\frac{5x}{2} + \frac{1}{4} = \frac{3}{4}$

25. $\frac{3x}{5} - \frac{x}{10} = \frac{2}{5}$ **26.** $\frac{4x}{3} - \frac{5}{6} = \frac{1}{6}$

27. $\frac{x+1}{3} + \frac{1}{6} = \frac{5}{6}$ **28.** $\frac{x-2}{4} - \frac{3x}{4} = \frac{3}{4}$

29. $\frac{x-2}{3} = \frac{x+5}{4}$ **30.** $\frac{2x-3}{3} = \frac{x+1}{2}$

2B - WORKSHEET: Solving Linear Equations

For 1 – 12, solve for the variable.

1. $6x = 12$	**2.** $4 + x = 12$
3. $2x + 3 = 9$	**4.** $3n + 2 = 5n - 3$
5. $3(2n - 5) = 4 - 3n$	**6.** $3x - 2 = -2(4x - 3)$
7. $3 - 4(5 - 4n) = 10$	**8.** $2 - 5x = 3 - 2(4 - x)$
9. $-7x + 4 = 3 - 5(4 - 2x)$	**10.** $\frac{3x}{4} = \frac{2}{3}$
11. $\frac{5x}{2} = \frac{4}{7}$	**12.** $\frac{8x}{3} = \frac{5}{4}$

For 13 – 18, solve for the indicated variable

13. $3x - 7y = 5;$ *solve for x*	**14.** $2xy = z;$ *solve for y*
15. $ax + by = c;$ *solve for y*	**16.** $\frac{2a}{b} = z;$ *solve for a*
17. $\frac{2z}{3} = \frac{5y}{x};$ *solve for z*	**18.** $2xy = 10st;$ *solve for y*

For 19 - 26, first simplify by expressing without fractions, then solve for x.

19. $\frac{3x}{5} + \frac{1}{5} = \frac{2}{5}$	**20.** $\frac{2x}{3} + \frac{1}{6} = \frac{5}{6}$
21. $\frac{2x}{5} + \frac{3x}{10} = \frac{2}{5}$	**22.** $\frac{4x+1}{2} - \frac{5}{4} = \frac{1}{4}$
23. $\frac{2x+1}{3} + \frac{1}{6} = \frac{7}{12}$	**24.** $\frac{x-2}{4} - \frac{5x}{8} = \frac{3}{4}$
25. $\frac{x-2}{2} = \frac{x+6}{5}$	**26.** $\frac{4x-3}{5} = \frac{x+2}{3}$

Answers:

1. $x = 2$ **2.** $x = 8$ **3.** $x = 3$ **4.** $n = \frac{5}{2}$ **5.** $n = \frac{19}{9}$ **6.** $x = \frac{8}{11}$ **7.** $n = \frac{27}{16}$ **8.** $x = 1$

9. $x = \frac{21}{17}$ **10.** $x = \frac{8}{9}$ **11.** $x = \frac{8}{35}$ **12.** $x = \frac{15}{32}$ **13.** $x = \frac{7y+5}{3}$ **14.** $y = \frac{z}{2x}$ **15.** $y = \frac{c-ax}{b}$

16. $a = \frac{zb}{2}$ **17.** $z = \frac{15y}{2x}$ **18.** $y = \frac{5st}{x}$ **19.** $\frac{1}{3}$ **20.** 1 **21.** $\frac{4}{7}$ **22.** $\frac{1}{2}$ **23.** $\frac{1}{8}$ **24.** $-\frac{10}{3}$ **25.** $\frac{22}{3}$

26. $\frac{19}{7}$

2C - Setting Up Equations

Algebra is useful for solving problems. In order to apply algebra to real situations, we must translate verbal expressions into algebraic expressions.

The following chart gives some common verbal expressions and their corresponding mathematical expressions. In this chart variable n represents "a number".

Verbal expression	Mathematical expression
the sum of a number and 10 a number plus 10 a number increased by 10 10 more than a number	$n + 10$
the difference between a number and 10 a number decreased by 10 10 subtracted from a number a number less 10 take away 10 from a number 10 less than a number 10 fewer than a number	$n - 10$
the product of a number and 10 a number multiplied by 10 a number times 10	$10n$
the quotient of a number and 10	$\dfrac{n}{10}$
a number is equal to 10 a number is 10	$n = 10$

EXAMPLE 1: Translate each of the following into a mathematical expression. Let the variable n represent "a number".

A) the sum of 3 and a number

B) the product of 5 and a number

C) the difference between 7 and a number

D) the difference between a number and 4

E) the quotient of 4 and a number

F) the quotient of a number and 9

G) a number decreased by 7

H) a number less 3

I) 6 subtracted from a number

J) 5 less than a number

SOLUTION:

A) $3 + n$

B) $5n$

C) $7 - n$

D) $n - 4$

E) $\dfrac{4}{n}$

F) $\dfrac{n}{9}$

G) $n - 7$

H) $n - 3$

I) $n - 6$

J) $n - 5$ ∎

EXAMPLE 2: Translate the following into a mathematical expression. Let x represent "a number".

A) the sum of twice an number and 3

B) four less than 5 times a number

C) Five decreased by three times a number

D) the quotient of 6 times a number and 5

E) twice the sum of a number and 10

F) Six more than the product of a number and 9

SOLUTION:

A) $2x + 3$

C) $5 - 3x$

E) $2(x + 10)$

B) $5x - 4$

D) $\frac{6x}{5}$

F) $9x + 6$ ∎

EXAMPLE 3: Translate the following sentence into a mathematical equation. Let n represent "a number".

A) A number increased by 3 is 14.

B) Twice a number increased by 7 is 15.

C) Three times a number decreased by 7 is 10 more than twice a number.

D) Twice the sum of a number and 5 is 2 less than 3 times the number.

E) Five less than 4 times a number is 10 more than the number.

SOLUTION:

A) $n + 3 = 14$

B) $2n + 7 = 15$

C) $3n - 7 = 10 + 2n$

D) $2(n + 5) = 3n - 2$

E) $4n - 5 = n + 10$ ∎

2C - EXERCISES

For 1 - 14 , translate the phrase into a mathematical expression. Let x represent "a number".

1. the sum of a number and 10

2. a number increased by 5

3. 6 more than a number

4. the difference between a 8 and a number

5. 7 less than a number

6. a number less 7

7. 6 subtracted from a number

8. twice a number

9. half a number

10. 5 less than twice a number

11. 9 more than 3 times a number

12. 4 times a number decrease by 7

13. twice the sum of a number and 6

14. two less than three times a number

For 15 - 20, translate the sentence into a mathematical equation. Let n represent "a number".

15. 10 more than a number is equal to 3.

16. A number less 7 is 4 more than twice the number.

17. 8 less than 3 times a number is two less than twice the number.

18. 7 more than three times a number is the number less 4.

19. 8 less than three times a number is two more than the number.

20. twice the sum of a number and 5 is 3 less than 7 times the number.

2C – WORKSHEET: Setting up equations

For 1-10, translate the phrase into an algebraic expression

1.	The sum of x and y.
2.	3 more than the sum of a and b
3.	16 less than the sum of x and y
4.	a multiplied by the sum of x and y
5.	a divided by x
6.	The sum of x and y multiplied by 5
7.	a less x
8.	5 less than x
9.	10 more than three times x
10.	4 less than 6 times x

For 11-14, translate the phrase into an algebraic equation. Let x represent a number.

11.	Ten added to twice a number is equal to five.
12.	Five subtracted from three times a number is equal to seven.
13.	Thirty is twelve less than three times a number.
14.	Three times the sum of a number and ten, is five.

For 15 – 19, translate the phrase into an algebraic equation and solve.

15.	Twice a number and seven is equal to eleven.
1.	Five less than four times a number is nineteen.
17.	Twice the sum of a number and ten is three times the number .

18.	The product of five and a number is three less than twice the number.
19.	Seven more than five times a number is six less than the number.

Answers:

1. $x + y$ **2.** $a + b + 3$ **3.** $x + y - 16$ **4.** $a(x + y)$ **5.** a/x **6.** $5(x + y)$ **7.** $a - x$ **8.** $x - 5$

9. $3x + 10$ **10.** $6x - 4$ **11.** $10 + 2x = 5$ **12.** $3x - 5 = 7$ **13.** $30 = 3x - 12$

14. $3(x + 10) = 5$ **15.** $x = 2$ **16.** $x = 6$ **17.** $x = 20$ **18.** $x = -1$ **19.** $x = -\dfrac{13}{4}$

2D - Inequalities

The following are **inequality symbols**: $<, >, \leq, and \geq$.

$2 < 3$ is read "2 is less than 3".

$2 \leq 3$ is read "2 is less than or equal to 3".

$3 > 2$ is read "3 is greater than 2".

$3 \geq 2$ is read "3 is greater than or equal to 2".

The inequality: $x < 2$ represents the numbers to the left of 2 on the number line:

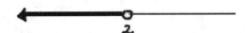   "x less than 2 "

The inequality: $x \leq 2$ represents the numbers to the left of 2 and equal to 2 on the number line:

   "x less than or equal to 2 "

The inequality: $x > -5$ represents the numbers to the right of -5 on the number line:

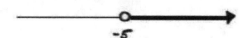   "x greater than -5 "

The inequality: $x \geq -5$ represents the numbers to the right of -5 and equal to -5 on the number line:

   "x greater than or equal to -5 "

Recall that a **linear expression** in one variable x is an expression that can be simplified to the form: $ax + b$, where a and b are constants.

A **linear inequality** is two linear expressions that are related by one of the inequality symbols. The following are examples of linear inequalities:

$2x + 3 < 5x - 2$

$3(x - 5) + 4 \geq 2 - 7(2x + 1)$

We solve linear inequalities in the same manner that we solve linear equations with one exception: **_when we multiply or divide both sides of the inequality by a negative number the direction of the inequality changes_**. For example,

The inequality: $-2 < 5$ is a true inequality. If we multiply both sides by -3 we have:

$(-3)(-2) > (-3)(5)$ *Notice that when we multiplied by -3 we changed the direction of the*
$6 > -15$ *inequality symbol so that the result is a true inequality.*

EXAMPLE 1: Solve for x.

A)

$$2(3x - 2) > 2x + 10$$
$$6x - 4 \quad > 2x + 10$$
$$\underline{-2x \qquad - 2x}$$
$$4x - 4 \quad > \qquad 10$$
$$\underline{\quad +4 \qquad\qquad + 4}$$
$$\frac{4x}{4} \qquad > \quad \frac{14}{4}$$
$$x \qquad\qquad > \quad \frac{7}{2}$$

B)

$$3(4 - 2x) \leq -2(3 + x)$$
$$12 - 6x \leq -6 - 2x$$
$$\underline{\quad +2x \qquad + 2x}$$
$$12 - 4x \leq -6$$
$$\underline{-12 \qquad - 12}$$
$$-4x \leq -18$$
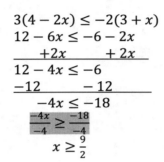
$$x \geq \frac{9}{2}$$

Notice that the direction of the inequality changed since we divided both sides by -4.

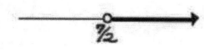

2D – EXERCISES

For $1 - 12$, Solve for x. Graph the solution on the number line.

1. $3x \geq 12$

2. $4x + 1 < 2x - 5$

3. $3 + 5(2x + 1) > 4(x + 3)$

4. $2 + 3(x - 5) < 2x + 1$

5. $-5x < 10$

6. $-\frac{3}{2}x \geq 12$

7. $-5x + 3 \leq 3x - 2$

8. $-10x + 4 > -8x + 10$

9. $3 - 2(5x - 3) > 7x + 1$

10. $2 - 3(2x + 1) < 3x + 4$

11. $-3(x - 4) + 3 \geq 5x + 7(3 - x)$

12. $-3(5 - x) + 7 \leq 5 - 3(x + 2)$

2D – WORKSHEET: Solve for x and graph the solution

1. $5x < 20$	**2.** $-3x < 30$
3. $5x + 3 \geq 2x - 4$	**4.** $7x + 2 \leq 3x - 4$
5. $-10x + 1 > 2x + 5$	**6.** $3x - 7 > 5x + 3$
7. $5x + 3 < 2x - 9$	**8.** $3 - 2(x + 4) > x - 4$
9. $3 - 7(2x - 3) \geq -3x - 2(x + 5)$	**10.** $x - 3(x + 5) < 4x + 2$

Answers:

1. $x < 4$

2. $x > -10$

3. $x \geq -7/3$

4. $x \leq -3/2$

5. $x < -1/3$

6. $x < -5$

7. $x < -4$

8. $x < -1/3$

9. $x \leq 34/9$

10. $x > -17/6$

2 – Answers to Exercises

Section A

1. 37 **2.** -7 **3.** -2 **4.** 18
5. 18 **6.** 66 **7.** -17 **8.** -5
9. 17 **10.** 42 **11.** $30 - 8x$ **12.** $7 - 9x$
13. $25 - 20x$ **14.** $-1 + 6x$ **15.** $-48 + 20x$ **16.** $-14 - 8x$
17. $27x - 18$ **18.** $-15x + 6$ **19.** $13x + 6$ **20.** $23 - 32x$
21. -4 **22.** 6 **23.** 39 **24.** 61

25. 12 **26.** 63 **27.** $2x + 1$ **28.** $3x$
29. $4x - 5$ **30.** $13x - 8$

Section B

1. $x = 2$ **2.** $x = 5$ **3.** $x = 2$ **4.** $x = 3$
5. $n = \frac{10}{3}$ **6.** $n = -\frac{1}{11}$ **7.** $x = 17$ **8.** $x = \frac{8}{3}$
9. $n = \frac{11}{12}$ **10.** $x = -\frac{1}{6}$ **11.** $x = 15$ **12.** $x = \frac{3}{2}$
13. $x = \frac{10}{7}$ **14.** $x = \frac{10}{9}$ **15.** $y = \frac{z-2x}{3}$ **16.** $y = \frac{c-ax}{b}$
17. $x = \frac{z}{3y}$ **18.** $x = \frac{3z}{2y}$ **19.** $h = \frac{2A}{b}$ **20.** $x = \frac{3z+5y}{2}$
21. $w = \frac{P-2l}{2}$ **22.** $c = \frac{2ad}{7b}$ **23.** $\frac{1}{2}$ **24.** $\frac{1}{5}$
25. $\frac{4}{5}$ **26.** $\frac{3}{4}$ **27.** 1 **28.** $-\frac{5}{2}$
29. 23 **30.** 9

Section C

1. $x + 10$ **2.** $x + 5$ **3.** $x + 6$ **4.** $8 - x$
5. $x - 7$ **6.** $x - 7$ **7.** $x - 6$ **8.** $2x$
9. $\frac{x}{2}$ **10.** $2x - 5$ **11.** $3x + 9$ **12.** $4x - 7$
13. $2(x + 6)$ **14.** $3x - 2$ **15.** $n + 10 = 3$ **16.** $n - 7 = 2n + 4$
17. $3n - 8 = 2n - 2$ **18.** $3n + 7 = n - 4$ **19.** $3n - 8 = n + 2$ **20.** $2(n + 5) = 7n - 3$

Section D

1. $x \geq 4$;

2. $x < -3$;

3. $x > \frac{2}{3}$;

4. $x < 14$;

5. $x > -2$;

6. $x \leq -8$;

7. $x \geq \frac{5}{8}$;

8. $x < -3$;

9. $x < \frac{8}{17}$;

10. $x > -\frac{5}{9}$;

11. $x \leq -6$;

12. $x \leq \frac{7}{6}$;

CHAPTER 3: Exponents and Square Roots

3A – The Exponent Rules

Recall that, $2^5 = 2 \cdot 2 \cdot 2 \cdot 2 \cdot 2 = 32$.

In the expression 2^5, 2 is the **base** and 3 is the **exponent** or **power**. The expression 2^5 is read: "2 *to the 5th power*" or "2 *to the 5th* ". We also say that 2^5 is a ***power of 2*** (since the power 5 is applied to the base 2).

We can simplify the product of two integer expressions having the same base.

$$2^5 \cdot 2^3 \;=\; \underbrace{\left(2 \cdot 2 \cdot 2 \cdot 2 \cdot 2\right)}_{5 \text{ twos}} \;\times\; \underbrace{\left(2 \cdot 2 \cdot 2\right)}_{3 \text{ twos}} \;=\; \underbrace{\left(2 \cdot 2 \cdot 2 \cdot 2 \cdot 2 \cdot 2 \cdot 2 \cdot 2\right)}_{5 + 3 \text{ twos}} \;=\; 2^{5+3} \;=\; 2^8$$

> **Multiplication Rule:** $a^n a^m = a^{n+m}$

EXAMPLE 1: Multiply. $(2x^3 y^2)(5x^4 y^7)$

SOLUTION: $(2x^3 y^2)(5x^4 y^7) = 10x^{3+4} y^{2+7} = 10y^7 x^9 \blacksquare$

We can simplify the quotient of two integer expressions having the same base.

$$\frac{2^5}{2^3} = \frac{2 \cdot 2 \cdot 2 \cdot 2 \cdot 2}{2 \cdot 2 \cdot 2} =$$

$$\frac{\overset{1}{\cancel{2}} \cdot \overset{1}{\cancel{2}} \cdot \overset{1}{\cancel{2}} \cdot 2 \cdot 2}{\underset{1}{\cancel{2}} \cdot \underset{1}{\cancel{2}} \cdot \underset{1}{\cancel{2}}} = 2^{5-3} = 2^2$$

> **Division rule:** $\dfrac{a^n}{a^m} = a^{n-m}, where\ a \neq 0$

EXAMPLE 2: Divide. **A)** $\frac{x^5 y^7}{x^2 y^3}$ **B)** $\frac{10 x^{10}}{5 x^5}$ **C)** $\frac{-3 x^7}{12 x^2}$

SOLUTION: A) $\frac{x^5 y^7}{x^2 y^3} = x^{5-2} y^{7-3} = x^3 y^4$ **B)** $\frac{10 x^{10}}{5 x^5} = 2 x^{10-5} = 2 x^5$ **C)** $\frac{-3 x^7}{12 x^2} = \frac{-x^{7-2}}{4} = \frac{-x^5}{4}$ ∎

We simplify the expression, $(2^5)^3$, as follows.

$$(2^5)^3 = (2^5)(2^5)(2^5) = 2^{5+5+5} = 2^{5\cdot3} = 2^{15}$$

Raising an Exponential Expression to a Power Rule:
$(a^n)^m = a^{n\cdot m}$

EXAMPLE 3: Simplify. **A)** $(2 x^4)^5$ **B)** $(-3 x^4 y^5)^2$

SOLUTION: A) $(2 x^4)^5 = 2^5 (x^4)^5 = 32 x^{4\cdot5} = 32 x^{20}$ **B)** $(-3 x^4 y^5)^2 = (-3)^2 (x^4)^2 (y^5)^2 = 9 x^8 y^{10}$ ∎

Let us simplify $\frac{2^3}{2^5}$ two ways. First,

$$\frac{2^3}{2^5} = \frac{2^3/2^3}{2^5/2^3} = \frac{1}{2^{5-3}} = \frac{1}{2^2}$$

Second, apply the division rule,

$$\frac{2^3}{2^5} = 2^{3-5} = 2^{-2} \quad \leftarrow \text{Notice the negative exponent .}$$

The following definition of negative integer exponents implies that the above two simplifications are equal.

Definition of a Negative Exponent: $a^{-n} = \frac{1}{a^n}$ $(a \neq 0)$

Notice that $\frac{1}{a^{-n}} = \frac{a^n}{1} = a^n$. For example,

$$\frac{1}{2^{-3}} = \frac{1}{\frac{1}{2^3}} = \frac{1\cdot2^3}{\frac{1}{2^3}\cdot2^3} = \frac{2^3}{1} = 2^3 .$$

EXAMPLE 4: Write the expression without negative exponents.

A) x^{-7} **B)** $\frac{1}{y^{-3}}$ **C)** $\frac{x^{-4}}{y^{-7}}$ **D)** $\frac{-2 x^{-5}}{y}$ **E)** $\frac{x^5}{4 y^{-3}}$ **F)** $3 x y^{-2}$

SOLUTIONS:

A) $x^{-7} = \frac{1}{x^7}$

B) $\frac{1}{y^{-3}} = y^3$

C) $\frac{x^{-4}}{y^{-7}} = \frac{1}{y^{-7}x^4} = \frac{y^7}{x^4}$

D) $\frac{-2x^{-5}}{y} = \frac{-2}{yx^5}$

E) $\frac{x^5}{4y^{-3}} = \frac{x^5y^3}{4}$

F) $3xy^{-2} = \frac{3x}{y^2}$ ∎

EXAMPLE 5: Divide. Write the answer without negative exponents.

A) $\frac{x^4}{x^{10}}$ **B)** $\frac{x^3y^8}{x^5y^2}$ **C)** $\frac{10x^5}{5x^{10}}$ **D)** $\frac{x^{-5}}{x^{10}}$ **E)** $\frac{-2x^{-3}}{x^7}$

SOLUTIONS:

A) $\frac{x^4}{x^{10}} = x^{4-10} = x^{-6} = \frac{1}{x^6}$

B) $\frac{x^3y^8}{x^5y^2} = x^{3-5}y^{8-2} = x^{-2}y^6 = \frac{y^6}{x^2}$

C) $\frac{10x^5}{5x^{10}} = \frac{2x^{5-10}}{1} = \frac{2x^{-5}}{1} = \frac{2}{x^5}$

D) $\frac{x^{-5}}{x^{10}} = \frac{1}{x^{10}x^5} = \frac{1}{x^{15}}$

E) $\frac{-2x^{-3}}{x^7} = \frac{-2}{x^7x^3} = \frac{-2}{x^{10}}$ ∎

If we apply the division of powers rule to the quotient $\frac{2^5}{2^5}$, we have

$$1 = \frac{2^5}{2^5} = 2^{5-5} = 2^0 \;.\; \leftarrow \text{Subtract exponents as in the division rule.}$$

The above example suggests the following rule.

The Zero Exponent Rule: $a^0 = 1, \;\; (a \neq 0)$.

EXAMPLE 6: Simplify. **A)** $\frac{3x^5}{x^5}$ **B)** $7x^5y^0z^4$

SOLUTION: A) $\frac{3x^5}{x^5} = 3x^0 = 3$ **B)** $7x^5y^0z^4 = 7x^5z^4$ ∎

It is convenient to have a rule for raising a fraction to a negative exponent.

$$\left(\frac{3}{4}\right)^{-2} = \frac{1}{\left(\frac{3}{4}\right)^2} = \frac{1}{\frac{9}{16}} = \frac{16}{9} = \left(\frac{4}{3}\right)^2$$

The above example suggests the rule,

Raising a Quotient to a Negative Exponent: $\left(\frac{a}{b}\right)^{-n} = \left(\frac{b}{a}\right)^{n}$

EXAMPLE 7: Simplify. **A)** $\left(\frac{x^3}{y^7}\right)^{-6} = \left(\frac{y^7}{x^3}\right)^{6} = \frac{y^{42}}{x^{18}}$ **B)** $\left(\frac{-2x^4}{3y^5}\right)^{-2} = \left(\frac{3y^5}{-2x^4}\right)^{2} = \frac{9y^{10}}{4x^8}$ ■

Here is a summary of the exponent rules:

The Exponent Rules
1. $a^n a^m = a^{n+m}$
2. $\frac{a^n}{a^m} = a^{n-m}, \ a \neq 0$
3. $(a^n)^m = a^{nm}$
4. $a^{-n} = \frac{1}{a^n}, a = 0$
5. $a^0 = 1, \ a \neq 0$
6. $\left(\frac{a}{b}\right)^{-n} = \left(\frac{b}{a}\right)^{n}, a \neq 0, b \neq 0$

3A – EXERCISES

For 1 – 14, simplify. Write answers without negative exponents.

1. $(2x^3)(5x^4)$
2. $(-3x^2y)(2x^4y^3)$
3. $(5x^5)(10x^{10})$
4. $(-4x^2y^5)(-3x^5y^2)$
5. $(5a^3b^4c^{10})(2a^7b^3c)$
6. $\frac{x^{10}}{x^3}$
7. $\frac{15x^6}{5x^2}$
8. $\frac{2x^3}{10x^2}$
9. $\frac{3x^5y^{10}}{12x^2y^3}$
10. $\frac{-5x^3y^4}{-10x^3y^2}$
11. $(x^3)^4$
12. $(2x^2y^5)^3$
13. $(-4a^5b^2)^2$
14. $(-3x^5)^3$

For 15 – 20, rewrite without negative exponents.

15. x^{-3}
16. $-5x^{-2}$
17. $-\frac{2y}{x^{-5}}$
18. $\frac{3}{-2x^{-4}}$
19. $\frac{x^{-3}}{y^{-5}}$
20. $\frac{x^2}{4y^{-3}}$

For 21 – 31, simplify. Write answer without negative exponents.

21. $\frac{x^3}{x^4}$
22. $\frac{5x^2y^{10}}{10x^5y^3}$
23. $\frac{-12x^{10}y^7}{-3x^2y^{10}}$
24. $\frac{x^{-5}}{x^2}$
25. $\frac{(x^3)^2}{x^4}$
26. $\frac{(x^4)^{-2}}{x^3}$
27. $\frac{3x^4y^5z^0}{9x^4y^2}$
28. $\frac{w^4x^3}{w^4x^{-5}}$
29. $\frac{x^5x^7}{(x^2)^4}$
30. $\left(\frac{4x^5}{y^3}\right)^{-2}$
31. $\left(\frac{y^3}{2x^4}\right)^{-3}$

3A – WORKSHEET: The Exponent Rules

For 1 – 14, simplify. Write answers without negative exponents.

1. $(3x^7)(7x^5)$	**2.** $(-7x^4y)(4x^8y^3)$	**3.** $(7x^7)(5x^5)$
4. $(-2x^3y^7)(-8x^4y^6)$	**5.** $(7a^2b^5c^{10})(3a^8b^2c)$	**6.** $\dfrac{x^8}{x^2}$
7. $\dfrac{10x^5}{5x^2}$	**8.** $\dfrac{2x^9}{16x^2}$	**9.** $\dfrac{3x^8y^{10}}{21x^2y^2}$
10. $\dfrac{-3x^5y^8}{-12x^5y^2}$	**11.** $(x^2)^7$	**12.** $(3x^4y^5)^3$
13. $(-5a^7b^2)^2$	**14.** $(-2x^4)^3$	

For 15 – 20, rewrite without negative exponents.

15. x^{-5}	**16.** $-7x^{-3}$	**17.** $-\dfrac{7y}{x^{-4}}$	**18.** $\dfrac{7}{-2x^{-5}}$	**19.** $\dfrac{x^{-8}}{y^{-4}}$	**20.** $\dfrac{x^4}{6y^{-5}}$

For 21 – 31, simplify. Write answer without negative exponents.

21. $\dfrac{x^{10}}{x^4}$	**22.** $\dfrac{5x^2y^8}{25x^6y^3}$	**23.** $\dfrac{-15x^{15}y^7}{-3x^2y^8}$	**24.** $\dfrac{x^{-3}}{x^2}$
25. $\dfrac{(x^5)^2}{x^3}$	**26.** $\dfrac{(x^3)^{-2}}{x^4}$	**27.** $\dfrac{3x^4y^9z^0}{6x^4y^2}$	**28.** $\dfrac{w^5x^3}{w^5x^{-4}}$
29. $\dfrac{x^4x^8}{(x^2)^3}$	**30.** $\left(\dfrac{5x^3}{y^7}\right)^{-2}$	**31.** $\left(\dfrac{y^4}{2x^5}\right)^{-3}$	

Answers:

1. $21x^{12}$ 2. $-28x^{12}y^4$ 3. $35x^{12}$

4. $16x^7y^{13}$ 5. $21a^{10}b^7c^{11}$ 6. x^6

7. $2x^3$ 8. $\dfrac{x^7}{8}$ 9. $\dfrac{x^6y^8}{7}$

10. $\dfrac{y^6}{4}$ 11. x^{14} 12. $27x^{12}y^{15}$

13. $25a^{14}b^4$ 14. $-8x^{12}$

15. $\dfrac{1}{x^5}$ 16. $-\dfrac{7}{x^3}$ 17. $-7yx^4$ 18. $-\dfrac{7x^5}{2}$ 19. $\dfrac{y^4}{x^8}$ 20. $\dfrac{x^4y^5}{6}$

21. x^6 22. $\dfrac{y^5}{5x^4}$ 23. $\dfrac{5x^{13}}{y}$ 24. $\dfrac{1}{x^5}$

25. x^7 26. $\dfrac{1}{x^{10}}$ 27. $\dfrac{y^7}{2}$ 28. x^7

29. x^6 30. $\dfrac{y^{14}}{25x^6}$ 31. $\dfrac{8x^{15}}{y^{12}}$

3B - Square Roots

The number 5 is the square root of 25 since $25 = 5 \cdot 5$. In general x is the square root of y if

$$y = x \cdot x = x^2 .$$

Notice that the square root of -25 is not a real number since there is no real number x such that $-25 = x^2$. In fact negative numbers do not have (real) square roots.

We write the square root of y as: $\sqrt{y}$. By definition, $(\sqrt{y})^2 = y$.

A square root such as $\sqrt{y}$ is called a **radical**; y is the **radicand**.

EXAMPLE 1: Find the square roots.

A) $\sqrt{9}$

B) $\sqrt{16}$

C) $\sqrt{64}$

D) $\sqrt{100}$

E) $\sqrt{36}$

F) $\sqrt{144}$

SOLUTION: A) 3 **B)** 4 **C)** 8 **D)** 10 **E)** 6 **F)** 12 ■

The square roots of most positive numbers are *not* whole numbers. For example, $\sqrt{26}$ is *approximately* equal to 5.1 ; ($(5.1)^2 = 26.01$).

We multiply two radicals according to the following rule.

The product rule: $\sqrt{a}\sqrt{b} = \sqrt{ab}$

EXAMPLE 2: Find the product and simplify.

A) $\sqrt{3}\sqrt{2}$

B) $(2\sqrt{3})(5\sqrt{2})$

C) $\sqrt{3}\sqrt{3}$

D) $\sqrt{2}(5 - 3\sqrt{2})$

E) $\sqrt{8}\sqrt{2}$

F) $(4\sqrt{3})(2\sqrt{12})$

G) $\sqrt{493}\sqrt{493}$

H) $\sqrt{3}(2\sqrt{7} - 5\sqrt{3})$

SOLUTIONS:

A) $\sqrt{3}\sqrt{2} = \sqrt{3 \cdot 2} = \sqrt{6}$

B) $(2\sqrt{3})(5\sqrt{2}) = 10\sqrt{6}$

C) $\sqrt{3}\sqrt{3} = (\sqrt{3})^2 = 3$

D) $\sqrt{2}(5 - 3\sqrt{2}) = 5\sqrt{2} - 3 \cdot 2 = 5\sqrt{2} - 6$

E) $\sqrt{8}\sqrt{2} = \sqrt{8 \cdot 2} = \sqrt{16} = 4$

F) $(4\sqrt{3})(2\sqrt{12}) = 8\sqrt{36} = 8 \cdot 6 = 48$

G) $\sqrt{493}\sqrt{493} = (\sqrt{493})^2 = 493$

H) $\sqrt{3}(2\sqrt{7} - 5\sqrt{3}) = 2\sqrt{21} - 15$ ■

We divide two radicals according to the following rule.

The quotient rule: $\dfrac{\sqrt{a}}{\sqrt{b}} = \sqrt{\dfrac{a}{b}}$

EXAMPLE 3: Find the quotient and simplify.

A) $\dfrac{\sqrt{12}}{\sqrt{3}}$

B) $\dfrac{\sqrt{2}}{\sqrt{50}}$

C) $\dfrac{5\sqrt{18}}{10\sqrt{2}}$

D) $\dfrac{\sqrt{5}\sqrt{15}}{\sqrt{3}}$

E) $\dfrac{\sqrt{3}\sqrt{21}}{\sqrt{7}}$

F) $\dfrac{\sqrt{5}}{\sqrt{11}\sqrt{55}}$

SOLUTION:

A) $\dfrac{\sqrt{12}}{\sqrt{3}} = \sqrt{\dfrac{12}{3}} = \sqrt{4} = 2$

B) $\dfrac{\sqrt{2}}{\sqrt{50}} = \sqrt{\dfrac{2}{50}} = \sqrt{\dfrac{1}{25}} = \dfrac{1}{5}$

C) $\dfrac{5\sqrt{18}}{10\sqrt{2}} = \dfrac{\sqrt{9}}{2} = \dfrac{3}{2}$

D) $\dfrac{\sqrt{5}\sqrt{15}}{\sqrt{3}} = \sqrt{\dfrac{5\cdot15}{3}} = \sqrt{25} = 5$

E) $\dfrac{\sqrt{3}\sqrt{21}}{\sqrt{7}} = \sqrt{\dfrac{3\cdot21}{7}} = \sqrt{9} = 3$

F) $\dfrac{\sqrt{5}}{\sqrt{11}\sqrt{55}} = \dfrac{1}{\sqrt{11}\sqrt{11}} = \dfrac{1}{11}$ ∎

We use the product rule to simplify square roots. For example,

$$\sqrt{50} = \sqrt{25 \cdot 2} = \sqrt{25} \cdot \sqrt{2} = 5\sqrt{2} \ .$$

EXAMPLE 4: Simplify the square roots.

A) $\sqrt{8}$
B) $\sqrt{32}$
C) $\sqrt{75}$

D) $\sqrt{12}$
E) $5\sqrt{27}$
F) $2\sqrt{48}$

SOLUTION:

A) $\sqrt{8} = \sqrt{4}\sqrt{2} = 2\sqrt{2}$
B) $\sqrt{32} = \sqrt{16}\sqrt{2} = 4\sqrt{2}$
C) $\sqrt{75} = \sqrt{25}\sqrt{3} = 5\sqrt{3}$

D) $\sqrt{12} = \sqrt{4}\sqrt{3} = 2\sqrt{3}$
E) $5\sqrt{27} = 5\sqrt{9}\sqrt{3} = 15\sqrt{3}$
F) $2\sqrt{48} = 2\sqrt{16}\sqrt{3} = 8\sqrt{3}$ ∎

Square roots with the same radicand are **like radicals.** Like radicals can be combined. For example, the radicals $4\sqrt{5}$ and $3\sqrt{5}$ are like radicals, and $4\sqrt{5} + 3\sqrt{5} = 7\sqrt{5}$.

The following radicals can be combined *after simplifying the first radical.*

$$\sqrt{50} + 3\sqrt{2} = \sqrt{25 \cdot 2} + 3\sqrt{2} = 5\sqrt{2} + 3\sqrt{2} = \boxed{8\sqrt{2}}$$

EXAMPLE 5: Combine the like radicals. If necessary, simplify first.

A) $3\sqrt{2} + \sqrt{25} + 7\sqrt{2} - 10$

B) $5\sqrt{6} - 3\sqrt{7} + 2\sqrt{6} + 5\sqrt{28}$

C) $3\sqrt{75} - 5\sqrt{3}$

D) $2\sqrt{12} - 5\sqrt{27}$

SOLUTION:

A) $3\sqrt{2} + \sqrt{25} + 7\sqrt{2} - 10 =$
$10\sqrt{2} + 5 - 10 =$
$10\sqrt{2} - 5.$

B) $5\sqrt{6} - 3\sqrt{7} + 2\sqrt{6} + 5\sqrt{28} =$
$7\sqrt{6} - 3\sqrt{7} + 5\sqrt{4}\sqrt{7} =$
$7\sqrt{6} - 3\sqrt{7} + 10\sqrt{7} = 7\sqrt{6} + 7\sqrt{7}$

C) $3\sqrt{75} - 5\sqrt{3} =$
$3\sqrt{25}\sqrt{3} - 5\sqrt{3} =$
$15\sqrt{3} - 5\sqrt{3} =$
$10\sqrt{3}.$

D) $2\sqrt{12} - 5\sqrt{27} =$
$2\sqrt{4}\sqrt{3} - 5\sqrt{9}\sqrt{3} =$
$4\sqrt{3} - 15\sqrt{3} =$
$-11\sqrt{3}$ ∎

EXAMPLE 6: Simplify completely.

A) $\dfrac{5\sqrt{6}\sqrt{21}}{\sqrt{2}}$

B) $\dfrac{\sqrt{14}\sqrt{6}}{\sqrt{3}}$

C) $\dfrac{\sqrt{3}\sqrt{15}}{\sqrt{5}}$

D) $\dfrac{\sqrt{6}}{\sqrt{21}\sqrt{14}}$

SOLUTION:

A) $\dfrac{5\sqrt{6}\sqrt{21}}{\sqrt{2}} = 5\sqrt{\dfrac{6 \cdot 21}{2}} = 5\sqrt{3 \cdot 21} = \sqrt{9 \cdot 7} =$
$15\sqrt{7}$

B) $\dfrac{\sqrt{14}\sqrt{6}}{\sqrt{3}} = \sqrt{\dfrac{14 \cdot 6}{3}} = \sqrt{14 \cdot 2} = \sqrt{4 \cdot 7} = 2\sqrt{7}$

C) $\dfrac{\sqrt{3}\sqrt{15}}{\sqrt{5}} = \sqrt{\dfrac{3 \cdot 15}{5}} = \sqrt{9} = 3$

D) $\dfrac{\sqrt{6}}{\sqrt{21}\sqrt{14}} = \sqrt{\dfrac{6}{21 \cdot 14}} = \sqrt{\dfrac{1}{49}} = \dfrac{1}{7}$ ∎

We can rewrite the radical expression, $\dfrac{1}{\sqrt{2}}$, by multiplying the numerator and denominator by $\sqrt{2}$.

$$\dfrac{1}{\sqrt{2}} \cdot \dfrac{\sqrt{2}}{\sqrt{2}} = \boxed{\dfrac{\sqrt{2}}{2}}$$

The result is a quotient whose denominator does not contain a radical. This process is called **rationalizing the denominator.** (The denominator of the result is a rational number.)

EXAMPLE 7: Rationalize the denominators.

A) $\dfrac{5}{2\sqrt{3}}$

B) $\dfrac{2\sqrt{3}}{\sqrt{6}}$

C) $\dfrac{5-\sqrt{7}}{\sqrt{7}}$

D) $\dfrac{6+2\sqrt{3}}{5\sqrt{3}}$

SOLUTION:

A) $\dfrac{5}{2\sqrt{3}} \cdot \dfrac{\sqrt{3}}{\sqrt{3}} = \dfrac{5\sqrt{3}}{2 \cdot 3} = \dfrac{5\sqrt{3}}{6}$

B) $\dfrac{2\sqrt{3}}{\sqrt{6}} \cdot \dfrac{\sqrt{6}}{\sqrt{6}} = \dfrac{2\sqrt{18}}{6} = \dfrac{6\sqrt{2}}{6} = \sqrt{2}$

C) $\dfrac{5-\sqrt{7}}{\sqrt{7}} \cdot \dfrac{\sqrt{7}}{\sqrt{7}} = \dfrac{5\sqrt{7}-7}{7}$

D) $\dfrac{6+2\sqrt{3}}{5\sqrt{3}} \cdot \dfrac{\sqrt{3}}{\sqrt{3}} = \dfrac{6\sqrt{3}+2\cdot3}{5\cdot3} = \dfrac{6\sqrt{3}+6}{15} = \dfrac{2\sqrt{3}+2}{5}$ ∎

The Pythagorean Theorem

A **right triangle** is a triangle that contains an angle of 90°; a **right angle**. The three sides of a right triangle are called the **legs** and the **hypotenuse**, where the hypotenuse is the longest side (the side opposite the right angle).

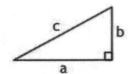 Sides **a** and **b** are legs, and side **c** is the hypotenuse.

The three sides of a right triangle are related according to the relationship:

$a^2 + b^2 = c^2$, known as the **Pythagorean Theorem**.

EXAMPLE 8: Find the missing side of the right triangle

A)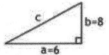

B)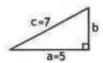

SOLUTION:

A) $c^2 = 6^2 + 8^2 = 36 + 64 = 100$
$c = 10$

B) $7^2 = 5^2 + b^2$
$49 = 25 + b^2$
$24 = b^2$
$\sqrt{24} = b$
$b = 2\sqrt{6}$ ∎

3B - EXERCISES

For 1 - 4 , find the square root.

1. $\sqrt{36}$

2. $\sqrt{121}$

3. $\sqrt{225}$

4. $\sqrt{10000}$

5. The quantity $\sqrt{250}$ is between which two integers ?

6. Is $4 - \sqrt{17}$ positive or negative?

For 7 - 18 , perform the indicated operation.

7. $\sqrt{18}\sqrt{2}$

8. $\sqrt{3}\sqrt{27}$

9. $(5\sqrt{7})(3\sqrt{7})$

10. $(2\sqrt{45})(3\sqrt{5})$

11. $(5\sqrt{3})^2$

12. $\sqrt{795}\sqrt{795}$

13. $\dfrac{\sqrt{50}}{\sqrt{2}}$

14. $\dfrac{\sqrt{48}}{\sqrt{3}}$

15. $\sqrt{100-64}$

16. $(2\sqrt{5})(4\sqrt{3})$

18. $\sqrt{25} + \sqrt{16}$

17. $\sqrt{\dfrac{25}{36}}$

For 19 - 24 , simplify the square root.

19. $\sqrt{75}$

20. $\sqrt{32}$

21. $\sqrt{48}$

22. $\sqrt{128}$

23. $\sqrt{96}$

24. $\sqrt{45}$

For 25 - 32 , combine.

25. $\sqrt{5} + 3\sqrt{5}$

26. $\sqrt{36} + 5\sqrt{7} - 10 - 2\sqrt{7}$

27. $7\sqrt{5} - 2\sqrt{45}$

28. $3\sqrt{50} + 2\sqrt{18}$

29. $\sqrt{27} - 2\sqrt{3}$

30. $5\sqrt{12} - 2\sqrt{48}$

31. $2\sqrt{8} + 5\sqrt{32}$

32. $4\sqrt{98} - 3\sqrt{2}$

For 33 - 36 , multiply

33. $\sqrt{3}(5 - \sqrt{3})$

34. $2\sqrt{5}(3 - 4\sqrt{5})$

35. $3\sqrt{6}(\sqrt{3} + \sqrt{6})$

36. $5\sqrt{2}(6 + 3\sqrt{2})$

For 37 - 40, simplify completely

37. $\dfrac{5\sqrt{3}\sqrt{6}}{\sqrt{2}}$

38. $\dfrac{\sqrt{12}}{3\sqrt{6}\sqrt{18}}$

39. $\dfrac{2\sqrt{5}\sqrt{15}}{\sqrt{12}}$

40. $\dfrac{\sqrt{14}\sqrt{18}}{\sqrt{7}}$

For 41 – 44, find the missing side.

41.

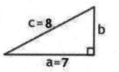
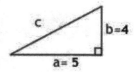

42.

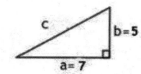
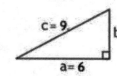

43.

44.

3B – WORKSHEET: Square Roots

A.

1. $\sqrt{64} =$ **2.** $\sqrt{144} =$

3. $\sqrt{169} =$ **4.** $\sqrt{196} =$

5. The quantity $\sqrt{105}$ is between which two integers?   **6.** The quantity $\sqrt{175}$ is between which two integers?

7. $\sqrt{3}\sqrt{12} =$ **8.** $\sqrt{8}\sqrt{2} =$

9. $\dfrac{\sqrt{32}}{\sqrt{2}} =$ **10.** $\dfrac{\sqrt{45}}{\sqrt{5}} =$

11. $\sqrt{5}\sqrt{5} =$ **12.** $\sqrt{347}\sqrt{347} =$

13. $\left(3\sqrt{5}\right)^2 =$ **14.** $\left(2\sqrt{5}\right)\left(3\sqrt{5}\right) =$

15. $\left(3\sqrt{12}\right)\left(2\sqrt{3}\right) =$ **16.** $\sqrt{16} + \sqrt{9} =$

17. $\sqrt{100 - 36} =$ **18.** $\sqrt{100} - \sqrt{36} =$

19. $\sqrt{\dfrac{16}{25}} =$ **20.** $\sqrt{\dfrac{49}{64}} =$

21. $\left(3\sqrt{5}\right)\left(2\sqrt{7}\right) =$ **22.** $\left(5\sqrt{2}\right)\left(3\sqrt{7}\right) =$

B. Simplify the square root.

1. $\sqrt{50} =$ **2.** $\sqrt{8} =$

3. $\sqrt{27} =$ **4.** $7\sqrt{45} =$

5. $\sqrt{288} =$ **6.** $\sqrt{108} =$

7. $\sqrt{243} =$ **8.** $\sqrt{128} =$

C. Combine.

1. $\sqrt{2} + 3\sqrt{2} =$ **2.** $\sqrt{25} - 5\sqrt{7} + 10 + 3\sqrt{7} =$

3. $\sqrt{20} + 3\sqrt{5} =$ **4.** $5\sqrt{98} + 3\sqrt{2} =$

5. $\sqrt{75} + \sqrt{27} =$ **6.** $4\sqrt{45} - 3\sqrt{20} =$

7. $3\sqrt{48} - 2\sqrt{27} =$ **8.** $3\sqrt{50} + \sqrt{32} =$

D. Multiply

1. $3\sqrt{5}\left(4 + 2\sqrt{5}\right) =$ **2.** $5\sqrt{2}\left(8 - 3\sqrt{7}\right) =$

3. $3\sqrt{7}\left(2 - \sqrt{2}\right) =$ **4.** $\sqrt{8}\left(2 + \sqrt{2}\right) =$

E. Simplify completely

1. $\dfrac{\sqrt{5}\sqrt{35}}{\sqrt{7}} =$ **2.** $\dfrac{\sqrt{6}\sqrt{30}}{\sqrt{5}} =$

3. $\dfrac{2\sqrt{35}\sqrt{21}}{\sqrt{3}} =$ **4.** $\dfrac{\sqrt{5}}{2\sqrt{3}\sqrt{15}} =$

F. Find the missing side

1.

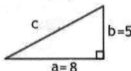

2.

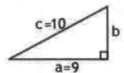

3.

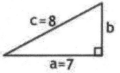

4.

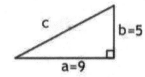

Answers: A. 1. 8 **2.** 12 **3.** 13 **4.** 14 **5.** 10 and 11 **6.** 13 and 14 **7.** 6 **8.** 4 **9.** 4 **10.** 3 **11.** 5 **12.** 347 **13.** 45 **14.** 30 **15.** 36 **16.** 7 **17.** 8 **18.** 4 **19.** 4/5 **20.** 7/8 **21.** $6\sqrt{35}$ **22.** $15\sqrt{14}$

B. 1. $5\sqrt{2}$ **2.** $2\sqrt{2}$ **3.** $3\sqrt{3}$ **4.** $21\sqrt{5}$ **5.** $12\sqrt{2}$ **6.** $6\sqrt{3}$ **7.** $9\sqrt{3}$ **8.** $8\sqrt{2}$

C. 1. $4\sqrt{2}$ **2.** $15 - 2\sqrt{7}$ **3.** $5\sqrt{5}$ **4.** $38\sqrt{2}$ **5.** $8\sqrt{3}$ **6.** $6\sqrt{5}$ **7.** $6\sqrt{3}$ **8.** $19\sqrt{2}$

D. 1. $30 + 12\sqrt{5}$ **2.** $40\sqrt{2} - 15\sqrt{14}$ **3.** $6\sqrt{7} - 3\sqrt{14}$ **4.** $4 + 4\sqrt{2}$

E. 1. 5 **2.** 6 **3.** $14\sqrt{5}$ **4.** $\frac{1}{6}$

F. 1. $\sqrt{15}$ **2.** $\sqrt{106}$ **3.** $\sqrt{89}$ **4.** $\sqrt{19}$

3 – Answers to Exercises

Section A

1. $10x^7$
2. $-6x^6y^4$
3. $50x^{15}$
4. $12x^7y^7$
5. $10a^{10}b^7c^{11}$
6. x^7
7. $3x^4$
8. $\frac{x}{5}$
9. $\frac{x^3y^7}{4}$
10. $\frac{y^2}{2}$
11. x^{12}
12. $8x^6y^{15}$
13. $16a^{10}b^4$
14. $-27x^{15}$
15. $\frac{1}{x^3}$
16. $-\frac{5}{x^2}$
17. $-2yx^5$
18. $-\frac{3x^4}{2}$
19. $\frac{y^5}{x^3}$
20. $\frac{x^2y^3}{4}$
21. $\frac{1}{x}$
22. $\frac{y^7}{2x^3}$
23. $\frac{4x^8}{y^3}$
24. $\frac{1}{x^7}$
25. x^2
26. $\frac{1}{x^{11}}$
27. $\frac{y^3}{3}$
28. x^8
29. x^4
30. $\frac{y^6}{16x^{10}}$
31. $\frac{8x^{12}}{y^9}$

Section B

1. 6
2. 11
3. 15
4. 100
5. 15 and 16
6. Negative
7. 6
8. 9
9. 105
10. 90
11. 75
12. 795
13. 5
14. 4
15. 6
16. $8\sqrt{15}$
17. $\frac{5}{6}$
18. 9
19. $5\sqrt{3}$
20. $4\sqrt{2}$
21. $4\sqrt{3}$
22. $8\sqrt{2}$
23. $4\sqrt{6}$
24. $3\sqrt{5}$
25. $4\sqrt{5}$
26. $-4 + 3\sqrt{7}$
27. $\sqrt{5}$
28. $21\sqrt{2}$
29. $\sqrt{3}$
30. $2\sqrt{3}$
31. $24\sqrt{2}$
32. $25\sqrt{2}$
33. $5\sqrt{3} - 3$
34. $6\sqrt{5} - 40$
35. $9\sqrt{2} + 18$
36. $30\sqrt{2} + 30$
37. 15
38. $\frac{1}{9}$
39. 5
40. 6
41. $\sqrt{15}$
42. $\sqrt{74}$
43. $\sqrt{41}$
44. $3\sqrt{5}$

CHAPTER 4: Polynomials

4A – Addition and Subtraction of Polynomials

A **monomial** is the product of variables and numbers. The following are examples of monomials:

$$3x^2, \qquad -5x^4y, \qquad 7x$$

A **polynomial** is the sum of monomials. The following are examples of polynomials (with one variable):

$$2x + 3, \qquad 3x^2 - 5x + 6, \qquad 4x^3 - 6x^2 - 2$$

Each monomial within a polynomial is called a **term**. The number in front of each term is called the **coefficient** of the term.

The **degree** of a polynomial (containing only one variable) is the largest exponent to which the variable appears. The degrees of the above three polynomials are 1, 2, and 3.

The degree 2 polynomial illustrated below has three terms.

$$\begin{array}{ccccc} 5x^2 & - & 7x & - & 4 \\ \uparrow & & \uparrow & & \uparrow \\ \text{term} & & \text{term} & & \text{term} \end{array}$$

with coefficients: $5, -7 \ and -4$

EXAMPLE 1: Evaluate the polynomial at the given value.

A) $5 - 3x; \ x = -2$ **B)** $-x^2 + 2x - 5; \ x = -3$

SOLUTION:
A) $5 - 3(-2) = 5 + 6 = 11$

B) $-(-3)^2 + 2(-3) - 5 = -(9) - 6 - 5 = -20$ ∎

Like terms have the same variables raised to the same exponents.   The following are examples of like terms.

Like terms:

$5x$	and	$-2x$
$-3x^2y$	and	$4x^2y$
$5xy$	and	$6yx$

We add or subtract polynomials by *combining like terms*. For example, $5x^2 + 3x^2 = 8x^2$. Notice that we combine like terms by adding the coefficients. We justify this process using the distributive law: $5x^2 + 3x^2 = (5 + 3)x^2 = 8x^2$.

EXAMPLE 2: Combine the like terms: $-3x^2 - 5x + 6 - 4x^2 + 7x - 9 = -7x^2 + 2x - 3$ ∎

EXAMPLE 3: $(-3x^4 - 5x^2 + 7) + (5x^4 - 4x^3 + 7x^2 - 2) =$

$-3x^4 - 5x^2 + 7 + 5x^4 - 4x^3 + 7x^2 - 2 = 2x^4 - 4x^3 + 2x^2 + 5$ ∎

We add polynomials enclosed in parenthesis by removing parenthesis and combining like terms.

EXAMPLE 4: Subtract. **A)** $(5x + 7) - (2x + 1)$ **B)** $(2x - 3) - (5x - 4)$

C) $(2x^2 - 4x + 5) - (5x^2 - 2x - 3)$

SOLUTION:

A) We subtract $5x - 2x$ and $7 - 1$ to obtain $3x + 6$. That is,

$(5x + 3) - (2x + 1) = 5x + 3 - 2x - 1 = 3x + 2$. *When we open parenthesis we subtract each term in the second polynomial.*

B) $(2x - 3) - (5x - 4) = 2x - 3 - 5x + 4 = -3x + 1$ *Notice that when we open parenthesis the signs of the second polynomial change.*

C) $(2x^2 - 4x + 5) - (5x^2 - 2x - 3) = 2x^2 - 4x + 5 - 5x^2 + 2x + 3 = -3x^2 - 2x + 8$

∎

4A – EXERCISES

For 1 – 4, evaluate the polynomial at the given value.

1. $2 - 7x; x = 5$
2. $3 - 4x; x = -2$
3. $-2x^2 - 5x + 3; \ x = -3$
4. $3x^2 - 4x - 2; \ x = -2$

For 5 – 14, perform the indicated operation.

5. $(7x + 3) + (5x + 2)$
6. $(3 - 7x) + (-4 - 3x)$
7. $(-5x^2 - 3x + 2) + (-7x^2 - 4x + 7)$
8. $(10x + 3) - (5x + 2)$
9. $(5x - 2) - (3x - 5)$
10. $(-2x^2 + 3x - 6) - (5x^2 - 2x + 3)$
11. $(5x^3 - 4x^2 + 3x - 2) + (7x - 3x^3 - 5)$
12. $(4x^2 - 3x + 2) - (-3x^2 + 2x - 5)$
13. $(10x^2 - 5x - 3) - (8x^2 + 2x - 5)$
14. $(3x^2 - 4x + 2) - (-4x^2 - 3x + 1)$

4A – WORKSHEET: Addition and Subtraction of Polynomials

For 1 – 4, evaluate the polynomial at the given value.

1. $5 - 3x; x = 2$	**2.** $7 - 4x; x = -2$
3. $-4x^2 - 2x + 1; \ x = -3$	**4.** $4x^2 - 5x - 2; \ x = -2$

For 5 – 14, perform the indicated operation.

5. $(8x + 1) + (7x + 2)$	**6.** $(7 - 5x) + (-2 - 3x)$
7. $(-4x^2 - 2x + 3) + (-5x^2 - 2x + 8)$	**8.** $(12x + 4) - (5x + 3)$
9. $(7x - 4) - (5x - 2)$	**10.** $(-3x^2 + 2x - 5) - (4x^2 - 6x + 1)$
11. $(4x^3 - 7x^2 + 2x - 1) + (5x - 2x^2 - 6)$	**12.** $(-5x^2 - 3x + 8) - (-4x^2 + 3x - 2)$

13. $(12x^2 - 4x - 2) - (7x^2 + 6x - 8)$	**14.** $(5x^2 - 7x + 6) - (-3x^2 - 2x + 5)$

Answers:

1. -1   **2.** 15   **3.** -29   **4.** 24

5. $15x + 3$ **6.** $5 - 8x$ **7.** $-9x^2 - 4x + 11$

8. $7x + 1$ **9.** $2x - 2$ **10.** $-7x^2 + 8x - 6$

11. $4x^3 - 9x^2 + 7x - 7$ **12.** $-x^2 - 6x + 10$ **13.** $5x^2 - 10x + 6$

14. $8x^2 - 5x + 1$

4B – Multiplication and Division of Polynomials

We multiply a polynomial by a monomial using the distributive law.

$$5x^2(3x + 7) = 15x^3 + 35x^2$$

EXAMPLE 1: Multiply.   **A)** $3x^2(2x^2 - 4x + 5) = 6x^4 - 12x^3 + 15x^2$

B) $-4x^3(3x^2 - 5x - 2) = -12x^5 + 20x^4 + 8x^3$ ∎

A polynomial with two terms is called a **binomial**.

We multiply two binomials by forming the four products resulting from multiplying each term in the first binomial with each term in the second. Below we demonstrate the preferred order for multiplication of two binomials.

$$(2x + 3)(4x + 5) =$$

$(2x)(4x) +$	$(2x)(5) +$	$(3)(4x) +$	$(3)(5) =$	
$8x^2 +$	$10x +$	$12x +$	15	$= 8x^2 + 22x + 15$
First	**Outer**	**Inner**	**Last**	

We label the four products: **First, Outer, Inner,** and **Last**. The **First** product is formed from the first term in each binomial. The **Outer** product is formed from the outer two terms in the expression. The **Inner** product is formed from the inner two terms in the expression. The **Last** product is formed from the last terms in each binomial. We use the acronym **FOIL** to remember the preferred order.

EXAMPLE 2: Multiply.   **A)** $(5x + 2)(3x + 5) = 15x^2 + 25x + 6x + 10 = 15x^2 + 31x + 10$

B) $(3x - 2)(7x + 5) = 21x^2 + 15x - 14x - 10 = 21x^2 + x - 10$ ∎

The pair of expressions, $A + B$ and $A - B$ are called **conjugates**. For example, $5x - 2$ and $5x + 2$ are conjugates. The product of conjugates is a binomial (two terms). For example,

$$(5x + 2)(5x - 2) =$$
$$25x^2 - 10x + 10x - 4 =$$ ← *Notice that the outer plus inner products*
$$25x^2 - 4$$ *cancel.*

A **trinomial** is a polynomial with three terms. We multiply a binomial by a trinomial using the distributive law twice.   For example,

$$(2x \quad + 3) \quad (4x^2 \quad +5x \quad +1) \quad =$$

$$\begin{array}{llll} 8x^3 & +10x^2 & +2x & & \leftarrow \text{Multiply} \quad 2x(4x^2 + 5x + 1) \\ & 12x^2 & +15x & +3 & \leftarrow \text{Multiply} \quad 3(4x^2 + 5x + 1)\ ; \text{line up like terms.} \\ \hline 8x^3 & +22x^2 & +17x & +3 & \leftarrow \text{Sum the two rows.} \end{array}$$

EXAMPLE 3: Multiply. $(3x - 4)(-2x^2 - 5x - 7) =$

SOLUTION:

$$\begin{array}{llll} -6x^3 & -15x^2 & -21x & \\ & 8x^2 & +20x & +28 \\ \hline -6x^3 & -7x^2 & -x & +28 \end{array}$$

∎

We divide a polynomial by a monomial dividing the monomial into each term of the polynomial. For example,

$$\frac{15x^4 - 5x^3}{5x^3} = \frac{15x^4}{5x^3} - \frac{5x^3}{5x^3} = 3x - 1$$

EXAMPLE 4: Divide. $\frac{21x^4 - 15x^3 + 3x^2}{-3x^2} = \frac{21x^4}{-3x^2} + \frac{-15x^3}{-3x^2} + \frac{3x^2}{-3x^2} = -7x^2 + 5x - 1$ ∎

4B – EXERCISES

1. $2x(3x + 5)$

2. $3x^2(5x^3 - 2x^2 + 4)$

3. $-5x^3(2x^2 - 3x + 4)$

4. $-6x^2(5x^3 - 2x^2 - 3x + 2)$

5. $(7x + 2)(3x + 5)$

6. $(3x - 2)(2x - 5)$

7. $(5x + 2)(2x - 3)$

8. $(2x + 3)(3x^2 - 4x + 2)$

9. $(3x - 2)(5x^2 - 3x - 2)$

10. $\frac{6x^5 - 4x^4 + 8x^3}{2x^3}$

11. $\frac{12x^5 + 3x^2}{3x^2}$

12. $\frac{12x^5 + 9x^2 - 3x}{-3x}$

13. $\frac{10x^5 - 15x^3 + 5x^2}{-5x^2}$

14. $(7x + 3)(7x - 3)$

4B – WORKSHEET: Multiplication and Division of Polynomials

1. $5x(2x+3)$	**2.** $4x^2(6x^3-3x^2+2)$
3. $-3x^3(4x^2-2x+7)$	**4.** $-5x^2(4x^3-3x^2-7x+2)$
5. $(5x+3)(4x+5)$	**6.** $(2x-5)(7x-3)$
7. $(6x+3)(8x-5)$	**8.** $(3x+2)(4x^2-2x+5)$
9. $(5x-2)(3x^2-2x-1)$	**10.** $\dfrac{10x^5-20x^4+15x^3}{5x^3}$
11. $\dfrac{12x^5+4x^2}{4x^2}$	**12.** $\dfrac{15x^5+12x^2-3x}{-3x}$

13. $\dfrac{14x^5 - 21x^3 + 7x^2}{-7x^2}$	14. $(5x + 7)(5x - 7)$

Answers:

1. $10x^2 + 15x$
2. $24x^5 - 12x^4 + 8x^2$
3. $-12x^5 + 6x^4 - 21x^3$
4. $-20x^5 + 15x^4 + 35x^3 - 10x^2$
5. $20x^2 + 37x + 15$
6. $14x^2 - 41x + 15$
7. $48x^2 - 6x - 15$
8. $12x^3 + 2x^2 + 11x + 10$
9. $15x^3 - 16x^2 - x + 2$
10. $2x^2 - 4x + 3$
11. $3x^3 + 1$
12. $-5x^4 - 4x + 1$
13. $-2x^3 + 3x - 1$
14. $25x^2 - 49$

4 – Answers to Exercises

Section A

1.	-33	**2.**	11	**3.**	0	**4.**	18

5. $12x + 5$ **6.** $-1 - 10x$ **7.** $-12x^2 - 7x + 9$

8. $5x + 1$ **9.** $2x + 3$ **10.** $-7x^2 + 5x - 9$

11. $2x^3 - 4x^2 + 10x - 7$ **12.** $7x^2 - 5x + 7$ **13.** $2x^2 - 7x + 1$

14. $7x^2 - x + 1$

Section B

1. $6x^2 + 10x$ **2.** $15x^5 - 6x^4 + 12x^2$

3. $-10x^5 + 15x^4 - 20x^3$ **4.** $-30x^5 + 12x^4 + 18x^3 - 12x^2$

5. $21x^2 + 41x + 10$ **6.** $6x^2 - 19x + 10$

7. $10x^2 - 11x - 6$ **8.** $6x^3 + x^2 - 8x + 6$

9. $15x^3 - 19x^2 + 4$ **10.** $3x^2 - 2x + 4$

11. $4x^3 + 1$ **12.** $-4x^4 - 3x + 1$

13. $-2x^3 + 3x - 1$ **14.** $49x^2 - 9$

5 – Factoring Polynomials

5A – Greatest Common Factors and Difference of Two Squares

Greatest Common Factors

When numbers or algebraic expressions are multiplied, they are **factors** of the result. The result is the **product**.

$$2 \quad \cdot \quad 3 \quad = \quad 6 \qquad\qquad 7x \cdot \quad (5x - 2) \quad = \quad 35x^2 - 14x$$

| ↑ | ↑ | ↑ | | ↑ | ↑ | ↑ |
| factor | factor | product | | factor | factor | product |

Factoring is a type of division where both the quotient and the divisor are unknown.

Division $\qquad\qquad\qquad\qquad\qquad$ Factoring

$$\text{divisor} \rightarrow \frac{x^2 + x}{x} = x + 1 \quad \leftarrow \text{quotient} \qquad x^2 + x = \quad x \cdot \quad (x + 1)$$

$$\qquad\qquad\qquad\qquad\qquad\qquad\qquad\qquad\qquad\qquad \uparrow \qquad \uparrow$$
$$\qquad\qquad\qquad\qquad\qquad\qquad\qquad\qquad\text{divisor} \quad \text{quotient}$$

EXAMPLE 1: Divide, then factor. $\frac{15x^3 + 25x^2}{5x^2} = 3x + 5$ and $15x^3 + 25x^2 = 5x^2(3x + 5)$ ∎

The expression $5x^2$ is a factor of both $15x^3$ and $25x^2$. We say that $5x^2$ is a **common factor** of the terms of polynomial $15x^3 + 25x^2$.

The terms of polynomial $36x^4 + 12x^3$ have many common factors such as $6x$, $6x^2, 12x^2$, and $12x^3$. The **greatest common factor** is $12x^3$.

EXAMPLE 2: Find the greatest common factor of the terms of polynomial $20x^2y^4 + 30x^3y^3$.

SOLUTION: The largest divisor of both 20 and 30 is 10. Expression x^2 is the highest power of x that divides both x^2 and x^3. Expression y^3 is the highest power of y that divides both y^4 and y^3. So $10x^2y^3$ is the greatest common factor. ∎

EXAMPLE 3: Factor. $12a^5b^3 - 18a^2b^4c$

SOLUTION: $12a^5b^3 - 18a^2b^4c = 6a^2b^3(2a^3 - 3bc)$; We can check our answer by multiplying.

Check: $6a^2b^3(2a^3 - 3bc) = 12a^5b^3 - 18a^2b^4c$ ∎

Factoring a Difference of Two Squares

We saw that the product of the sum of two numbers and the difference of two numbers is the difference of the squares of the two numbers. That is,

$$(A + B)(A - B) = A^2 - AB + BA - B^2 = A^2 - B^2.$$

The reverse of this equality is the formula for factoring a difference of two squares.

Difference of Two Squares Formula: $A^2 - B^2 = (A + B)(A - B)$

The expression $25x^2$ is a perfect square since $25x^2 = (5x)^2$. The expression $9x^{10}$ is also a perfect square as $9x^{10} = (3x^5)^2$. (x^n is a perfect square if n is even.) The following are examples of polynomials that are differences of two squares.

$$x^2 - 9, \quad 9x^2 - 25, \quad 16x^2 - 49y^2, \quad 9x^2y^2 - 4z^2, \quad x^4 - 25$$

If a polynomial with two terms is a difference of two squares then it can be factored as in the above formula.

EXAMPLE 4: Factor. **A)** $25x^2 - 9$ **B)** $36x^2 - 49y^2$ **C)** $9x^4 - 25y^2$

SOLUTION:

A) $25x^2 - 9 = (5x)^2 - (3)^2 = (5x + 3)(5x - 3)$
B) $36x^2 - 49y^2 = (6x)^2 - (7y)^2 = (6x + 7y)(6x - 7y)$
C) $9x^4 - 25y^2 = (3x^2)^2 - (5y)^2 = (3x^2 + 5y)(3x^2 - 5y)$ ∎

We factor polynomial $x^3 - 4x$ in two steps. First pull out the common factor, x, then factor the difference of two squares.

$$x^3 - 4x = x \cdot (x^2 - 4) = x \cdot \underbrace{(x + 2)(x - 2)}$$

Factored as a difference of two squares.

EXAMPLE 5: Factor *completely*. **A)** $10x^2 - 90$ **B)** $18x^5y^5 - 8x^3y^7$

SOLUTION:
A) $10x^2 - 90 = 10(x^2 - 9) = 10(x + 3)(x - 3)$
B) $18x^5y^5 - 8x^3y^7 = 2x^3y^5(9x^2 - 4y^2) = 2x^3y^5(3x + 2y)(3x - 2y)$ ∎

5A – EXERCISES

1. **A)** Divide. $\dfrac{25x^3+10x}{5x}$ **B)** Factor. $25x^3 + 10x$

2. **A)** Divide. $\dfrac{12x^3y+6x^2y^2}{6x^2y}$ **B)** Factor. $12x^3y + 6x^2y^2$

For 3 – 8, factor out the greatest common factor.

3. $5x^2 + 10x$
5. $12a^2b^3c - 18a^2b^4 + 6a^3b^3$
7. $50x^4y^5 + 40x^3y^7$

4. $18x^2 - 12x^3y$
6. $14x^2yz^3 - 21x^3y^2z^2$
8. $49x^2y^5z^3 - 14x^3y^4z^4$

For 9 – 12, factor as a difference of two squares.

9. $x^2 - 25$
11. $9x^2y^2 - 16z^2$

10. $4x^2 - 25$
12. $49x^2z^2 - 64a^2$

For 13 – 18, factor completely.

13. $10x^2 - 40$
15. $45x^3y - 20xy^3$
17. $175zx^2 - 28zy^2$

14. $2x^3 - 18x$
16. $40x^3y - 90xy^3$
18. $27yx^2 - 48y^3$

5A – WORKSHEET: Greatest Common Factors and Difference of Two Squares

1. A) Divide. $\dfrac{12x^3+18x}{6x}$ B) Factor. $12x^3 + 18x$

2. A) Divide. $\dfrac{25x^3y+5x^2y}{5x^2y}$ B) Factor. $25x^3y + 5x^2y$

For 3 – 8, factor out the greatest common factor.

3. $15x^2 + 25x$	4. $21x^3 - 14x^4y$
5. $5a^2b^3c - 15a^2b^4 + 10a^3b^3$	6. $18x^2yz^3 - 12x^3y^2z^2$
7. $300x^4y^5 + 500x^3y^7$	8. $36x^2y^6z^3 - 27x^3y^5z^4$

For 9 – 12, factor as a difference of two squares.

9. $x^2 - 49$	10. $9x^2 - 16$
11. $25x^2y^2 - 4z^2$	12. $64y^2z^2 - 9b^2$

For 13 – 18, factor completely.

13. $10x^2 - 160$	14. $3x^3 - 27x$
15. $27x^3y - 12xy^3$	16. $250x^3y - 90xy^3$
17. $32zx^2 - 18zy^2$	18. $75yx^2 - 48y^3$

Answers:

1. A) $2x^2 + 3$ B) $6x(2x^2 + 3)$
2. A) $5x + 1$ B) $5x^2y(5x + 1)$
3. $5x(3x + 5)$
4. $7x^3(3 - 2xy)$
5. $5a^2b^3(c - 3b + 2a)$
6. $6x^2yz^2(3z - 2xy)$
7. $100x^3y^5(3x + 5y^2)$
8. $9x^2y^5z^3(4y - 3xz)$
9. $(x + 7)(x - 7)$
10. $(3x + 4)(3x - 4)$
11. $(5xy - 2z)(5xy + 2z)$
12. $(8yz - 3b)(8yz + 3b)$
13. $10(x + 4)(x - 4)$
14. $3x(x + 3)(x - 3)$
15. $3yx(3x + 2y)(3x - 2y)$
16. $10xy(5x + 3y)(5x - 3y)$
17. $2z(4x + 3y)(4x - 3y)$
18. $3y(5x + 4y)(5x - 4y)$

5B – Factoring by Grouping and Factoring Quadratic Expressions

Factoring by Grouping

The polynomial $x(*) + 7(*)$ has two terms with a common factor of $(*)$.

$$x(*) + 7(*) = (*)(x + 7)$$

Similarly, the polynomial $x(2a + 1) + 7(2a + 1)$ has two terms with a common factor of $(2a + 1)$.

$$x(2a + 1) + 7(2a + 1) = (2a + 1)(x + 7)$$

Now, if we remove parenthesis, $x(2a + 1) + 7(2a + 1) = 2ax + x + 14a + 7$.

We factor $2ax + x + 14a + 7$ by reversing this process.

$$
\begin{aligned}
&\underline{2ax + x} \quad + \quad \underline{14a + 7} \qquad \leftarrow \text{ Group together the first two} \\
&\qquad\qquad\qquad\qquad\qquad\qquad\quad \text{terms and the last two terms.} \\
&= \quad x(2a + 1) \quad + \quad 7(2a + 1) \qquad \leftarrow \text{ Factor each group.} \\
&= \qquad\quad (2a + 1)(x + 7)
\end{aligned}
$$

The above is an example of **factoring by grouping**.

EXAMPLE 1: Factor by grouping.

A) $15ax - 10ay + 12bx - 8by = 5a(3x - 2y) + 4b(3x - 2y) = (3x - 2y)(5a + 4b)$

B) $6xa - 10xb - 15ya - 25yb = 2x(3a - 5b) - 5y(3a - 5b) = (3a - 5b)(2x - 5y)$ ∎

Factoring Quadratic Expressions

A quadratic is a polynomial of the form $ax^2 + bx + c, \ a \neq 0$, such as

$$
\begin{array}{ccccc}
3x^2 & - & 13x & - & 10 \\
\uparrow & & \uparrow & & \uparrow \\
a = 3 & & b = -13 & & c = -10
\end{array}
$$

The values 3, and -13 are the **coefficients** of x^2 and x. Value -10 is the **constant coefficient**.

Some quadratics can be factored by factoring out a greatest common factor.

$$5x^2 + 10x = 5x(x + 2)$$

Some quadratics can be factored as a difference of two squares.

$$9x^2 - 25 = (3x + 5)(3x - 5)$$

EXAMPLE 2: Factor. **A)** $8x^2 + 4x$ **B)** $9x^2 - 16$

SOLUTION:
A) $8x^2 + 4x = 4x(2x + 1)$; pull out greatest common factor
B) $9x^2 - 16 = (3x + 4)(3x - 4)$; factor as a difference of two squares   ■

The quadratic $3x^2 - 13x - 10 = (3x + 2)(x - 5)$ is the product of two binomials. The product $(3x + 2)(x - 5)$ is the factored form. Recall that we multiply binomials using the FOIL procedure. We factor $3x^3 - 13x - 10$ using a process called **reverse FOIL**.

The following example is designed to prepare for factoring quadratics using reverse FOIL.

EXAMPLE 3: Fill in the missing terms. It is best to do these mentally.

A) $(x + 5)(x + 3) = x^2 + 8x + [\ \ ]$ **B)** $(2x - 3)(x + 1) = [\ \ ] - x - 3$
C) $(y + 7)(y - 3) = y^2 + [\ \ ] - 21$ **D)** $([\ \ ] + 2)(x + 3) = x^2 + 5x + 6$
E) $(x + [\ \ ])(x + 5) = x^2 + 8x + 15$ **F)** $([\ \ ] + 7)([\ \ ] + 3) = x^2 + 10x + 21$
G) $([\ \ ] + 1)([\ \ ] + 2) = 3x^2 + 7x + 2$ **H)** $([\ \ ] + [\ \ ])(x + 2) = x^2 + 6x + 8$
I) $(5x - 3)(2x - 1) = 10x^2 + [\ \ ] + 3$ **J)** $(2x + 5)(3x - 7) = 6x^2 + [\ \ ] - 35$

SOLUTION:
A) 15   **B)** $2x^2$ **C)** $4y$ **D)** x **E)** 3
F) x, x **G)** $3x, x$ **H)** $x, 4$ **I)** $-11x$ **J)** x ■

EXAMPLE 4: Fill in the missing signs. It is best to do these mentally.

A) $(x\ \ 3)(x\ \ 2) = x^2 + 5x + 6$ **B)** $(x\ \ 3)(x\ \ 2) = x^2 - 5x + 6$
C) $(x\ \ 7)(x\ \ 3) = x^2 - 4x - 21$ **D)** $(2x\ \ 3)(3x\ \ 1) = 6x^2 - 7x - 3$
E) $(3x\ \ 1)(2x\ \ 3) = 6x^2 + 7x - 3$

SOLUTION: A) $+, +$ **B)** $-, -$ **C)** $-, +$ **D)** $-, +$ **E)** $-, +$ ■

Factoring Quadratics of the form $ax^2 + bx + c$, where $c > 0$:
If $c > 0$, then $ax^2 + bx + c = (* + *)(* + *)\ or\ (* - *)(* - *)$

For example,

$x^2 + 5x + 6 = (x + 3)(x + 2)$ and $x^2 - 10x + 21 = (x - 3)(x - 7)$.

EXAMPLE 5: Factor. **A)** $x^2 + 12x + 35$ **B)** $x^2 - 8x + 15$ **C)** $2x^2 - 7x + 5$ **D)** $3x^2 + 11x + 6$

SOLUTIONS:
A) $x^2 + 12x + 35 = (x + 5)(x + 7)$
B) $x^2 - 8x + 15 = (x - 5)(x - 3)$ Notice that if $c > 0$ and $b < 0$, then the signs are $(-)(-)$.
C) $2x^2 - 7x + 5 = (2x - 5)(x - 1)$
D) $3x^2 + 11x + 6 = (3x + 2)(x + 3)$ ■

Factoring Quadratics of the form $ax^2 + bx + c$, where $c < 0$:
If $c < 0$, then $ax^2 + bx + c = (* + *)(* - *)$ or $(* - *)(* + *)$

EXAMPLE 6: Factor. **A)** $x^2 - 2x - 15$ **B)** $x^2 + 5x - 14$ **C)** $2x^2 - x - 3$ **D)** $3x^2 + x - 2$

SOLUTIONS:
A) $x^2 - 2x - 15 = (x + 3)(x - 5)$
B) $x^2 + 5x - 14 = (x - 2)(x + 7)$
C) $2x^2 - x - 3 = (2x - 3)(x + 1)$ Always check that the O+I (FOIL) products sum to the middle
                                         term. In this case $-2x + 3x = -x$
D) $3x^2 + x - 2 = (3x - 2)(x + 1)$ O+I $= 3x - 2x = x$ ■

Factoring Quadratics by the Method of Grouping

The quadratic $3x^2 + 17x + 20$ can be written as $3x^2 + 12x + 5x + 20$. We can apply the factoring by grouping method.

$$3x^2 + 12x + 5x + 20 = 3x(x + 4) + 5(x + 4) = (x + 4)(3x + 5).$$

Notice that the middle term, $17x$, was split into $12x + 5x$. The product $(12)(5) = 60$ and the product of the first and last coefficients $(3)(20) = 60$. Number 60 is our **grouping number** and is used to split $17x$ into $12x + 5x$.

Factoring $ax^2 + bx + c$ by grouping
1. Find the grouping number: ac
2. Find two numbers whose sum is b and whose product is ac.
3. Rewrite the middle term, bx, as a sum using these two numbers.
4. Factor by grouping.

5B– EXERCISES

For $1 - 8$, factor by the method of grouping.

1. $2ax + 2a + 3bx + 3b$
2. $3ax + 9x + 10ay + 30y$
3. $6ax + 3a - 10bx - 5b$
4. $6by + 10cy + 9bz + 15cz$
5. $2x^3 + 10x^2 - 3x - 15$
6. $15x^3 - 10x^2 - 21x + 14$

7. $2ax + 5bx + 6a + 15b$

8. $15x^3 + 6x^2 + 35x + 14$

For 9 – 12, factor the quadratic by the method of greatest common factor or difference of two squares (whichever one applies).

9. $12x^2 + 18x$

10. $10x^2 - 30x$

11. $25x^2 - 64$

12. $4x^2 - 81$

13. $21x^2 - 14x$

14. $36x^2 - 49$

For 15 – 30, factor by the method of reverse FOIL (or grouping).

15. $x^2 + 5x + 6$

16. $x^2 - 12x + 35$

17. $x^2 + 8x + 12$

18. $2x^2 - 7x + 6$

19. $3x^2 + 8x + 5$

20. $7x^2 - 23x + 6$

21. $3x^2 - 19x + 6$

22. $5x^2 + 19x + 14$

23. $x^2 + 4x - 21$

24. $2x^2 + x - 15$

25. $3x^2 + x - 14$

26. $7x^2 - 34x - 5$

27. $5x^2 + 18x - 8$

28. $3x^2 + 5x - 12$

29. $2x^2 - 5x - 33$

30. $6x^2 - 23x - 55$

5B – WORKSHEET: Factoring by Grouping and Factoring Quadratic Expressions

For 1 – 8, factor by the method of grouping.

1. $5ax + 5a + 7bx + 7b$	**2.** $6ax + 4x + 15ay + 10y$
3. $6ax - 2a - 15bx + 5b$	**4.** $15by + 25cy + 6bz + 10cz$
5. $5x^3 + 10x^2 - 7x - 14$	**6.** $28x^3 - 8x^2 - 35x + 10$
7. $3ax + 7bx + 15a + 35b$	**8.** $10x^3 - 15x^2 - 14x + 21$

For 9 – 12, factor the quadratic by the method of greatest common factor or difference of two squares (whichever one applies).

9. $5x^2 + 15x$	**10.** $12x^2 - 18x$
11. $49x^2 - 64$	**12.** $25x^2 - 9$
13. $24x^2 - 12x$	**14.** $16x^2 - 9$

For 15 – 30, factor by the method of reverse FOIL (or grouping).

15. $x^2 + 8x + 15$	**16.** $x^2 - 9x + 14$
17. $x^2 + 10x + 16$	**18.** $2x^2 - 11x + 15$

19. $3x^2 + 10x + 7$	**20.** $5x^2 - 17x + 6$
21. $5x^2 - 31x + 6$	**22.** $3x^2 + 17x + 14$
23. $x^2 - 2x - 15$	**24.** $2x^2 + 7x - 15$
25. $3x^2 - x - 10$	**26.** $11x^2 - 54x - 5$
27. $7x^2 + 26x - 8$	**28.** $5x^2 + 11x - 12$
29. $3x^2 - 5x - 22$	**30.** $6x^2 - 7x - 55$

Answers:

1. $(x + 1)(5a + 7b)$
2. $(3a + 2)(2x + 5y)$
3. $(3x - 1)(2a - 5b)$
4. $(3b + 5c)(5y + 2z)$
5. $(x + 2)(5x^2 - 7)$
6. $(7x - 2)(4x^2 - 5)$
7. $(x + 5)(3a + 7b)$
8. $(2x - 3)(5x^2 - 7)$
9. $5x(x + 3)$
10. $6x(2x - 3)$
11. $(7x + 8)(7x - 8)$
12. $(5x + 3)(5x - 3)$
13. $12x(2x - 1)$
14. $(4x + 3)(4x - 3)$
15. $(x + 3)(x + 5)$
16. $(x - 7)(x - 2)$
17. $(x + 8)(x + 2)$
18. $(2x - 5)(x - 3)$
19. $(3x + 7)(x + 1)$
20. $(5x - 2)(x - 3)$
21. $(5x - 1)(x - 6)$
22. $(3x + 14)(x + 1)$
23. $(x + 3)(x - 5)$
24. $(2x - 3)(x + 5)$
25. $(3x + 5)(x - 2)$
26. $(11x + 1)(x - 5)$
27. $(7x - 2)(x + 4)$
28. $(5x - 4)(x + 3)$
29. $(3x - 11)(x + 2)$
30. $(3x - 11)(2x + 5)$

5C – Solving Quadratic Equations by Factoring

The following are examples of quadratic equations.

$$x^2 = 25 \qquad 2x^2 + 11x + 5 = 0 \qquad x(x - 1) = 0$$

The equation $ax^2 + bx + c = 0$ is quadratic equation written in **standard form** ($a \neq 0$).

Most quadratic equations have two solutions (or roots). For example, $x^2 = 25$ has solutions $x = 5$ and $x = -5$. One method of solving quadratic equations is the **factoring method**. It is based on the following rule.

The $A \cdot B = 0$ rule If $A \cdot B = 0$, then $A = 0$ or $B = 0$.	← If the product of two numbers is 0 then at least one of the numbers must be equal to 0.

The Factoring Method for Solving a Quadratic Equation
1. Put the quadratic equation in standard form $ax^2 + bx + c = 0$.
2. Factor $ax^2 + bx + c$ (if possible).
3. Set each factor equal to zero and solve.

EXAMPLE 1: Solve for x.   **A)**$x^2 - 10x + 21 = 0$ **B)**$10x^2 + 30x = 0$ **C)**$9x^2 - 25 = 0$

SOLUTIONS:

A) $x^2 - 10x + 21 = (x - 3)(x - 7) = 0$ The quadratic equation is in standard form. So we start by factoring (reverse FOIL).

$$x - 3 = 0 \qquad\qquad x - 7 = 0$$
$$x = 3 \qquad\qquad\quad x = 7$$

Set each factor equal to zero and solve. The two solutions are $x = 3$ and $x = 7$.

Check: $x = 3$; $(3)^2 - 10(3) + 21 =$
$9 - 30 + 21 = 0$.
$\quad x = 7$; $(7)^2 - 10(7) + 21$
$\qquad\qquad = 49 - 70 + 21 = 0$.

We can check our solutions by substituting into the original equation.

B) $10x^2 + 30x = 10x(x + 3) = 0$ The quadratic equation is in standard form. So we start by factoring (greatest common factor).

$$10x = 0 \qquad\qquad x + 3 = 0$$
$$x = 0 \qquad\qquad\quad x = -3$$

Set each factor equal to zero and solve. The two solutions are $x = 0$ and $x = -3$.

C) $9x^2 - 25 = (3x + 5)(3x - 5) = 0$ The quadratic equation is in standard form. So we start by factoring (difference of two squares).

$$3x + 5 = 0 \qquad\qquad 3x - 5 = 0$$
$$3x = -5 \qquad\qquad\quad 3x = 5$$
$$x = -\frac{5}{3} \qquad\qquad\quad x = \frac{5}{3}$$

Set each factor equal to zero and solve. The two solutions are $x = -\frac{5}{3}$ and $x = \frac{5}{3}$. ∎

EXAMPLE 2: Solve for x.   **A)**$2x^2 + 3x = 35$ **B)**$10x^2 = 160$

SOLUTIONS:

A) $2x^2 + 3x = 35$

$2x^2 + 3x - 35 = (2x - 7)(x + 5) = 0$

$$2x - 7 = 0 \qquad x + 5 = 0$$
$$2x = 7 \qquad x = -5$$
$$x = \frac{7}{2}$$

Put in standard form. Subtract 35 from both sides of the equation.
Factor.

Set each factor equal to zero and solve. The two solutions are $x = \frac{7}{2}$ and $x = -5$.

B)
$$10x^2 = 160$$
$$x^2 = 16$$
$$x^2 - 16 = 0$$
$$(x + 4)(x - 4) = 0$$

Divide both sides by 10 to simplify.
Put in standard form. Subtract 16 from both sides of the equation. Factor.

$$x + 4 = 0 \qquad x - 4 = 0$$
$$x = -4 \qquad x = 4$$

Set each factor equal to zero and solve. The two solutions are $x = -4$ and $x = 4$. ∎

5C – EXERCISES

For 1 – 12, solve for x.

1. $x^2 + 12x + 35 = 0$

2. $x^2 - 8x + 15 = 0$

3. $x^2 - x = 0$

4. $5x^2 + 10x = 0$

5. $4x^2 - 49 = 0$

6. $9x^2 - 16 = 0$

7. $2x^2 - 17x + 21 = 0$

8. $3x^2 + 13x - 10 = 0$

9. $7x^2 - 33x - 10 = 0$

10. $14x^2 + 29x - 15 = 0$

11. $3x^2 - 28x - 55 = 0$

12. $2x^2 - 13x + 6 = 0$

For 13 – 22, put in standard form and solve for x.

13. $12x^2 = 27$

14. $10x^2 = 490$

15. $50x^2 = 32$

16. $x^2 = -6x$

17. $15x^2 = 10x$

18. $x^2 - 3x = 28$

19. $2x^2 + 15 = 13x$

20. $5x^2 - 53x = 22$

21. $3x^2 + 17x = 6$

22. $7x^2 - 10 = 33x$

5C – WORKSHEET: Solving Quadratic Equations By Factoring

For 1 – 12, solve for x.

1. $x^2 + 10x + 21 = 0$	**2.** $x^2 - 7x + 10 = 0$
3. $x^2 - 4x = 0$	**4.** $7x^2 + 14x = 0$
5. $25x^2 - 9 = 0$	**6.** $4x^2 - 81 = 0$
7. $3x^2 - 13x + 14 = 0$	**8.** $2x^2 - x - 15 = 0$
9. $5x^2 - 3x - 14 = 0$	**10.** $14x^2 + 11x - 15 = 0$
11. $5x^2 - 52x - 33 = 0$	**12.** $3x^2 - 46x + 15 = 0$

For 13 – 22, put in standard form and solve for x.

13. $5x^2 = 45$	**14.** $10x^2 = 250$
15. $5x^2 = -15x$	**16.** $21x^2 = 35x$

17. $x^2 + 4x = 21$	**18.** $5x^2 - 6x = 8$
19. $3x^2 - 13x = -14$	**20.** $2x^2 - 17x = 55$
21. $7x^2 + 41x = 6$	**22.** $5x^2 - 6 = -13x$

Answers:

1. $-7, -3$	**2.** $5, 2$	**3.** $0, 4$	**4.** $0, -2$
5. $-\frac{3}{5}, \frac{3}{5}$	**6.** $-\frac{9}{2}, \frac{9}{2}$	**7.** $\frac{7}{3}, 2$	**8.** $-\frac{5}{2}, 3$
9. $-\frac{7}{5}, 2$	**10.** $-\frac{3}{2}, \frac{5}{7}$	**11.** $-\frac{3}{5}, 11$	**12.** $\frac{1}{3}, 15$
13. $3, -3$	**14.** $5, -5$	**15.** $0, -3$	**16.** $0, \frac{5}{3}$
17. $3, -7$	**18.** $-\frac{4}{5}, 2$	**19.** $\frac{7}{3}, 2$	**20.** $-\frac{5}{2}, 11$
21. $\frac{1}{7}, -6$	**22.** $\frac{2}{5}, -3$		

5 – Answers to Exercises

Section A

1. **A)** $5x^2 + 2$ **B)** $5x(5x^2 + 2)$
2. **A)** $2x + y$ **B)** $6x^2y(2x + y)$
3. $5x(x + 2)$
4. $6x^2(3 - 2xy)$
5. $6a^2b^3(2c - 3b + a)$
6. $7x^2yz^2(2z - 3xy)$
7. $10x^3y^5(5x + 4y^2)$
8. $7x^2y^4z^3(7y - 2xz)$
9. $(x + 5)(x - 5)$
10. $(2x + 5)(2x - 5)$
11. $(3xy + 4z)(3xy - 4z)$
12. $(7xz + 8a)(7xz - 8a)$
13. $10(x + 2)(x - 2)$
14. $2x(x + 3)(x - 3)$
15. $5xy(3x + 2y)(3x - 2y)$
16. $10xy(2x + 3y)(2x - 3y)$
17. $7z(5x + 2y)(5x - 2y)$
18. $3y(3x + 4y)(3x - 4y)$

Section B

1. $(x + 1)(2a + 3b)$
2. $(a + 3)(3x + 10y)$
3. $(2x + 1)(3a - 5b)$
4. $(3b + 5c)(2y + 3z)$
5. $(x + 5)(2x^3 - 3)$
6. $(3x - 2)(5x^2 - 7)$
7. $(x + 3)(2a + 5b)$
8. $(5x + 2)(3x^2 + 7)$
9. $6x(2x + 3)$
10. $10x(x - 3)$
11. $(5x + 8)(5x - 8)$
12. $(2x + 9)(2x - 9)$
13. $7x(3x - 2)$
14. $(6x - 7)(6x + 7)$
15. $(x + 2)(x + 3)$
16. $(x - 7)(x - 5)$
17. $(x + 6)(x + 2)$
18. $(2x - 3)(x - 2)$
19. $(3x + 5)(x + 1)$
20. $(7x - 2)(x - 3)$
21. $(3x - 1)(x - 6)$
22. $(5x + 14)(x + 1)$
23. $(x + 7)(x - 3)$
24. $(2x - 5)(x + 3)$
25. $(3x + 7)(x - 2)$
26. $(7x + 1)(x - 5)$
27. $(5x - 2)(x + 4)$
28. $(3x - 4)(x + 3)$
29. $(2x - 11)(x + 3)$
30. $(3x + 5)(2x - 11)$

Section C

1. $-5, -7$
2. $5, 3$
3. $0, 1$
4. $0, -2$
5. $-\frac{7}{2}, \frac{7}{2}$
6. $\frac{4}{3}, -\frac{4}{3}$
7. $\frac{3}{2}, 7$
8. $\frac{2}{3}, -5$
9. $-\frac{2}{7}, 5$
10. $\frac{3}{7}, -\frac{5}{2}$
11. $-\frac{5}{3}, 11$
12. $\frac{1}{2}, 6$
13. $-\frac{3}{2}, \frac{3}{2}$
14. $7, -7$
15. $\frac{4}{5}, -\frac{4}{5}$
16. $0, -6$
17. $0, \frac{2}{3}$
18. $7, -4$
19. $\frac{3}{2}, 5$
20. $-\frac{2}{5}, 11$
21. $\frac{1}{3}, -6$
22. $-\frac{2}{7}, 5$

6 – Graphs of Lines

6A – The Rectangular Coordinate System and Slopes

The **coordinate plane** is based on two number-lines, a horizontal and a vertical. The horizontal number-line is the $x - axis$ and the vertical is the $y - axis$. The intersection point is the **origin** and is located at 0 on both number-lines.

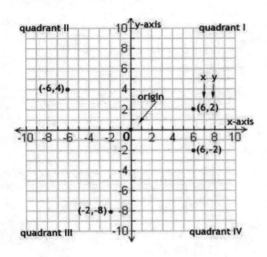

- **Points** on the coordinate plane are named by a pair of numbers. The first number is the $x - coordinate$ and it indicates how far to the right or left the point is from the origin. The second number is the $y - coordinate$ and it indicates how many units up or down the point is from the origin. The origin is $(0,0)$.

- A negative $x - coordinate$ indicates that the point is to the left of the origin, and a negative $y - coordinate$ indicates that the point is below the origin.

EXAMPLE 1: Plot the following points on the coordinate plane: $(3,4), (-6,1), (-8,-2), (2,-4)$.

SOLUTION:

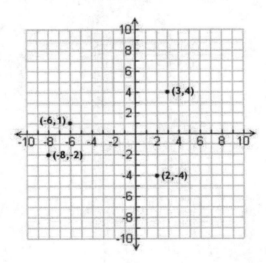

The incline of a line is measured by its **slope**. When viewed from left to right, a line with positive slope points up and a line with negative slope points down. Below are some graphs of lines and their slopes.

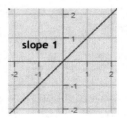 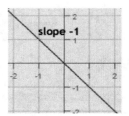  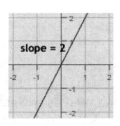 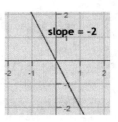

To motivate the definition of the slope of a line we first define the slope of a staircase. The slope of a staircase is measured by the quotient of the vertical dimension (rise) and the horizontal dimension (run). A typical staircase has

$$slope = \frac{6\ inches}{10\ inches} = \frac{3}{5}.$$

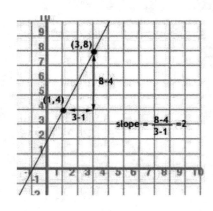

The slope of a line is similar to the slope of a staircase. It is a quotient of a vertical dimension and a horizontal dimension.

Given (x_1, y_1) and (x_2, y_2), two points on a line, the slope is defined as:

$$\textbf{Slope } = \frac{y_2 - y_1}{x_2 - x_1}$$

EXAMPLE 2: Find the slope of the line that goes through the given points.

A) $(2,3)\ and\ (7,8)$ **B)** $(-3,5)\ and\ (6,-4)$.

SOLUTION: A) $slope = \frac{8-3}{7-2} = \frac{5}{5} = 1$ **B)** $slope = \frac{-4-5}{6-(-3)} = \frac{-9}{9} = -1$ ∎

EXAMPLE 3: Graph a line given a point on the line and the slope of the line.

A) point $(2,1)$ and slope $\frac{3}{4}$.   **B)** point $(3,2)$ and slope -4.

SOLUTION:

A) First plot the point (2,1). Starting at point (2,1) count 3 units up and 4 units to the right. Plot the landing point. Connect the two points.

B) First plot point (3,2). Write the slope as the fraction, $\frac{-4}{1}$. Starting at (3,2) count 4 units *down* and 1 unit to the right. Plot the landing point. Connect the two points.

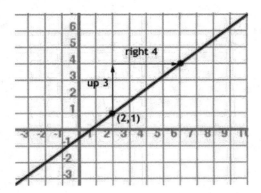

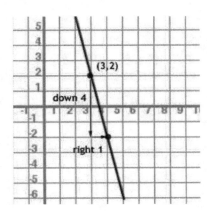

■

EXAMPLE 4: Determine the slope of the line from the graph.

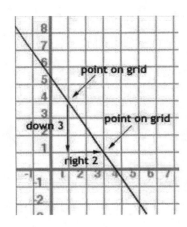

SOLUTION:

First find two points that fall on the grid. Starting at one of the points count the number of units up or down and then the number of units right or left to the second point. The slope is the quotient of the vertical over the horizontal. The two directions, *down and left*, are indicated by negative numbers (up and right are positive). So the slope of the line is: $\frac{-3}{2}$. ■

6A – EXERCISES

1. Plot the points: $(3,5), (-2,3), (-3,-4), (5,-2)$.

For 2 - 5 , find the slope of the line through the given points.

2. $(2,3)\,and\,(5,6)$ 3. $(-8,2)\,and\,(4,-6)$ 4. $(-4,-5)\,and\,(-2,3)$ 5. $(6,-9)\,and\,(-3,12)$

For $6-8$, plot the line through the given point and with the given slope.

6. Point (3,2), slope = $\frac{5}{6}$.

7. Point (1,5), slope = $\frac{-3}{2}$.

8. Point (3,1), slope = -2.

For 9 - 11, find the slope of the line given the graph.

9.

10.

11.

6A – WORKSHEET: The Rectangular Coordinate System and Slopes

1. Plot the points: $(2,4), (-3,2), (-5,-3), (7,-4)$.

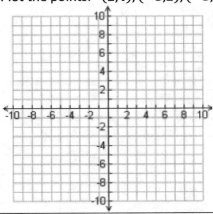

For 2 - 5 , find the slope of the line through the given points.

2. $(3,2) and (7,6)$	**3.** $(-6,3) and (5,-7)$	**4.** $(-2,-7) and (-4,2)$	**5.** $(8,-7) and (-4,10)$

For 6 – 8, plot the line through the given point and with the given slope.

6. Point $(3,1)$, slope $= \frac{4}{3}$.	**7.** Point $(5,1)$, slope $= \frac{-2}{3}$.	**8.** Point $(1,4)$, slope $= -3$.
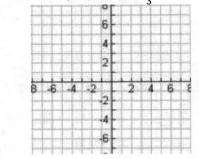	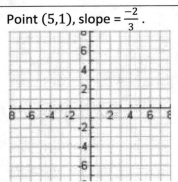	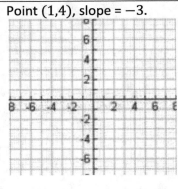

For 9 - 11, find the slope of the line given the graph.

9.	10.	11.
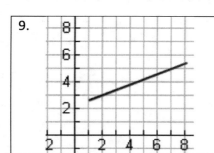		

Answers:

1.

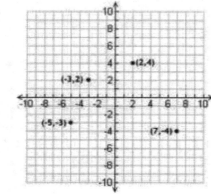

2.	1	3.	$-\dfrac{10}{11}$	4.	$\dfrac{9}{-2}$	5.	$\dfrac{17}{-12}$

6.	7.	8.
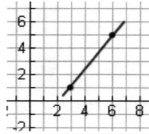	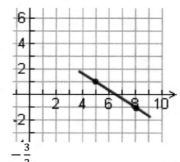	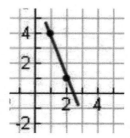

9.	$\dfrac{2}{5}$	10.	$-\dfrac{3}{2}$	11.	$\dfrac{1}{2}$

6B- The Equation of a Line

The equation $x + y = 10$ has solutions that are pairs of numbers, (x, y), that sum to 10.
For example,
$(5,5), (4,6), (6,4), (7,3), (9.9, 0.1),$ and $(-90, 100)$ are all solutions of $x + y = 10$. If we plot the solutions we see that they line up in a straight line. The equation $x + y = 10$ is the equation of the line pictured on the right.

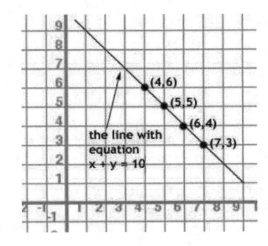

In section 6A we plotted lines given a point on the line and the slope of the line. We find the equation of a line from the same information.

Given point (x_1, y_1) and slope m, the equation of the corresponding line is given by:

$$y - y_1 = m(x - x_1) \leftarrow \textbf{the Point-Slope Line Equation}$$

Note: The point-slope line equation is derived from the slope formula.

EXAMPLE 1: Find the equation of the line through the given point with the given slope. Solve for y.

A) point $(2,3)$ and slope $m = 5$ **B)** point $(-3,4)$ and slope $m = -2$

SOLUTION:

A)
$$y - 3 = 5(x - 2)$$
$$y - 3 = 5x - 10$$
$$\underline{+3 \qquad + \ 3} \leftarrow solve\ for\ y$$
$$y \qquad = 5x - \ 7$$

$$y = 5x - 7$$

B)
$$y - 4 = -2(x - (-3))$$
$$y - 4 = -2(x + 3)$$
$$y - 4 = -2x - 6 \leftarrow solve\ for\ y$$
$$\underline{+4 = \qquad + 4}$$
$$y \qquad = -2x - 2 \quad \blacksquare$$

The $x - \textbf{\textit{intercept}}$ is the point where a line crosses the $x - axis$. The $y - \textbf{\textit{intercept}}$ is the point where a line crosses the $y - axis$.

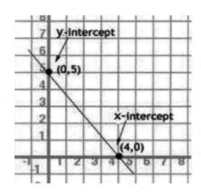

On the right we have a line with $x - intercept$ $(4,0)$ and $y - intercept$ $(0,5)$. Notice that the $y - coordinate$ of the $x - intercept$ is 0 and the $x - coordinate$ of the $y - intercept$ is 0.

Sometimes we say that the $y - intercept$ is 5.

Let us find the equation of the line with $y - intercept$ $(0,2)$ and slope 5. Using the point-slope line equation we have:

$$y - 2 = 5(x - 0)$$
$$y - 2 = 5x$$
$$+2 = \quad +2$$
$$y \quad = 5x + 2$$

← Notice that the slope 5 is the $x - coefficient$ and the $y - intercept$ 2 is the constant.

In general a line with equation $y = mx + b$ has slope m and $y - intercept$ b or $(0, b)$.

$y = mx + b$ ←	The **Slope-Intercept Line Equation**. The slope is m and the $y -$ intercept is b or $(0, b)$.

EXAMPLE 2: Find the slope and $y - intercept$ of the line with equation **A)** $y - 3x = 2$ **B)** $2y + 5x = 10$.

SOLUTION: **A)** $y - 3x = 2$ ← Put in form $y = mx + b$.
$$+3x \quad +3x$$
$$y \quad = 3x + 2$$ ← The slope is 3 and the $y - intercept$ is 2 or $(0,2)$.

B) $2y + 5x = 10$ ← Put in form $y = mx + b$.
$$-5x = -5x$$
$$2y \quad = -5x + 10$$
$$y \quad = -\frac{5}{2}x + 5$$ ← The slope is $-\frac{5}{2}$ and the $y - intercept$ is 5. ■

EXAMPLE 3: Using the $y = mx + b$ line equation, find the equation of the line with slope 5 and containing point $(-2,3)$.

SOLUTION: $y = mx + b$ ← Use the slope-intercept line equation. Replace m with the slope of 5.

$$y = 5x + b$$ ← Replace x with -2 and y with 3 and solve for b.
$$3 = 5(-2) + b$$
$$3 = -10 + b$$
$$13 = b$$

8
Chapter 6

$$y = 5x + 13 \qquad \leftarrow \text{The equation.} \qquad \blacksquare$$

EXAMPLE 4: Find the equation of the line through points $(1,4)$ and $(2,-3)$.

SOLUTION: $m = \dfrac{y_2 - y_1}{x_2 - x_1} = \dfrac{-3 - 4}{2 - 1} = \dfrac{-7}{1} = -7 \qquad \leftarrow$ Find the slope of the line.

$$y = -7x + b \qquad \leftarrow \text{Substitute point } (1,4) \text{ (or } (2,-3)\text{) and solve for } b.$$

$$4 = -7(1) + b$$
$$b = 11$$
$$y = -7x + 11 \qquad \leftarrow \text{The equation of the line through points } (1,4) \text{ and } (2,-3). \quad \blacksquare$$

EXAMPLE 5: Graph the line with equation $3y - 2x = -12$.

SOLUTION:
$$3y - 2x = -12 \qquad \leftarrow \text{Put into the form } y = mx + b.$$
$$3y = 2x - 12$$
$$y = \frac{2}{3}x - 4 \qquad \leftarrow \text{The slope is } \frac{2}{3} \text{ and the } y - intercept \text{ is } -4.$$

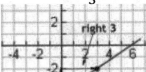

$\leftarrow$ Start at the $y - intercept$, $(0,-4)$. Use the slope of $\frac{2}{3}$ to find another point (go up 2 and go right 3). Connect the two points.

$\blacksquare$

Horizontal and vertical lines

A **horizontal line** has slope 0. The equation of a horizontal line is $y = mx + b$ with $m = 0$:

$$\boxed{y = b \qquad \leftarrow \text{The equation of a horizontal line.}}$$

EXAMPLE 6: Graph the horizontal line with equation $y = 5$.

SOLUTION: All points with $y - coordinate$ 5 are on the graph of line $y = 5$. For example , $(0,5)$ and $(-4,5)$ are on the graph. We will plot these points and draw the line.

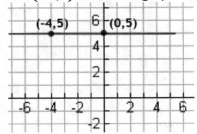

$\leftarrow$ The horizontal line $y = 5$.

$\blacksquare$

EXAMPLE 7: Find the equation of the horizontal line that goes through point $(17,-4)$.

SOLUTION: The equation is of the form $y = b$ where b is the $y - coordinate$ of any point on the line. Since point $(17, -4)$ is on the line, the equation is $y = -4$. ∎

A **vertical line** has an undefined (or infinite) slope. All points on the vertical line have the same $x - coordinate$. The equation of a vertical line is of the form:

$$x = a \quad \leftarrow \text{ The equation of a vertical line}$$

EXAMPLE 8: Plot the vertical line with equation $x = 5$.

SOLUTION: Any point with $x - coordinate$ 5 is on the line. Plot two points with $x - coordinate$ 5, such as $(5,0)$ and $(5,4)$. Draw the line containing the two points.

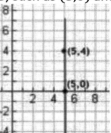

$\leftarrow$ The graph of line $x = 5$. ∎

EXAMPLE 9: Find the equation of the vertical line containing point $(17,18)$.

SOLUTION: The equation of a vertical line is of the form $x = a$. Value a is the $x - coordinate$ of any point on the line. So, the equation is $x = 17$. ∎

6B – EXERCISES

For $1 - 3$, find the equation of the line with the given slope and through the given point.

1. $m = 5; (2,3)$ **2.** $m = -2; (-1,4)$ **3.** $m = 7; (-4, -3)$

For $4 - 6$, find the equation of the line with the given slope and $y - intercept$.

4. $m = -2; (0,8)$ **5.** $m = \frac{2}{3}; (0, -7)$ **6.** $m = -\frac{5}{6}; (0,239)$

For $7 - 12$, find the slope and $y - intercept$ of the line with the given equation.

7. $y = 22x - 39$ **8.** $2y + 6x = 12$ **9.** $3y = -2x + 9$
10. $x - 5y - 20 = 0$ **11.** $15x + 5y = 4$ **12.** $7x - 3y = -12$

For 13 – 17, find the equation of the line through the given 2 points.

13. $(2,3), (1,7)$ **14.** $(-2,5), (-4,-3)$ **15.** $(1,9), (4,-3)$
16. $(0,3), (-1,-2)$ **17.** $(2,-4), (4,-2)$

For 18 – 20, draw the graph of the line with the given equation.

18. $y = 2x + 1$ **19.** $2y - 3x = -4$ **20.** $3y + 4x = 3$

21. Graph the horizontal line with equation $y = 2$.
22. Graph the vertical line with equation $x = -4$.
23. Find the equation of the horizontal line containing point $(-2,4)$.
24. Find the equation of the horizontal line containing point $(5,6)$.
25. Find the equation of the vertical line containing point $(7,8)$.
26. Find the equation of the vertical line containing point $(-3,7)$.

6B – WORKSHEET: The equation of the line

For 1 – 3, find the equation of the line with the given slope and through the given point.

1. $m = 3; (2,5)$	2. $m = -3; (-2,7)$	3. $m = 5; (-3,-2)$

For 4 – 6, find the equation of the line with the given slope and $y - intercept$.

4. $m = -4; (0,6)$	5. $m = \frac{4}{5}; (0,-3)$	6. $m = -\frac{4}{9}; (0,739)$

For 7 – 12, find the slope and $y - intercept$ of the line with the given equation.

7. $y = 13x - 77$	8. $3y + 6x = 12$	9. $4y = -2x + 12$
10. $x - 7y - 21 = 0$	11. $25x + 5y = 3$	12. $7x - 4y = -16$

For 13 – 17, find the equation of the line through the given 2 points.

13. $(2,5), (1,8)$	14. $(-2,7), (-4,-5)$	15. $(1,12), (4,-3)$
16. $(0,4), (-1,-3)$	17. $(-2,-6), (-4,-2)$	

For 18 – 20, draw the graph of the line with the given equation.

18. $y = \frac{2}{3}x + 1$	19. $2y + 3x = -4$	20. $3y + x = 6$

21.	Graph the horizontal line with equation $y = -3$.
22.	Graph the vertical line with equation $x = -5$.
23.	Find the equation of the horizontal line containing point $(-3,7)$.
24.	Find the equation of the horizontal line containing point $(4,13)$.
25.	Find the equation of the vertical line containing point $(20,30)$.
26.	Find the equation of the vertical line containing point $(-30,15)$.

Answers:

1. $y = 3x - 1$
2. $y = -3x + 1$
3. $y = 5x + 13$
4. $y = -4x + 6$
5. $y = \frac{4}{5}x - 3$
6. $y = -\frac{4}{9}x + 739$
7. $m = 13, (0, -77)$
8. $m = -2, (0,4)$
9. $m = -\frac{1}{2}, (0,3)$
10. $m = \frac{1}{7}, (0, -3)$
11. $m = -5, \left(0, \frac{3}{5}\right)$
12. $m = \frac{7}{4}, (0,4)$
13. $y = -3x + 11$
14. $y = 6x + 19$
15. $y = -5x + 17$
16. $y = 7x + 4$
17. $y = -2x - 10$
18.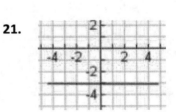

19.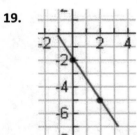

20.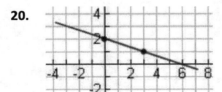

21.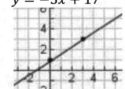

22.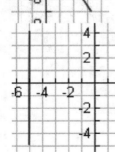

23. $y = 7$

24. $y = 13$

25. $x = 20$

26. $x = -30$

6 – Answers to Exercises

Section A

1.

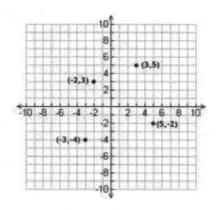

2. 1

3. $-\dfrac{2}{3}$

4. 4

5. $-\dfrac{7}{3}$

6.

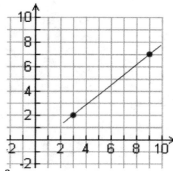

7.

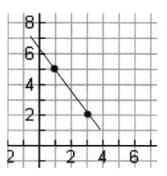

8.

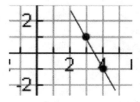

9. $\dfrac{3}{2}$

10. $-\dfrac{2}{3}$

11. -3

Section B

1. $y = 5x - 7$

2. $y = -2x + 2$

3. $y = 7x + 25$

4. $y = -2x + 8$

5. $y = \dfrac{2}{3}x - 7$

6. $y = -\dfrac{5}{6}x + 239$

7. $m = 22, (0, -39)$

8. $m = -3, (0,6)$

9. $m = -\dfrac{2}{3}, (0,3)$

10. $m = \dfrac{1}{5}, (0, -4)$

11. $m = -3, \left(0, \dfrac{4}{5}\right)$

12. $m = \dfrac{7}{3}, (0,4)$

13. $y = -4x + 11$

14. $y = 4x + 13$

15. $y = -4x + 13$

16. $y = 5x + 3$

17. $y = x - 6$

18.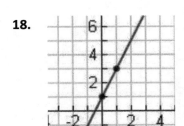

19.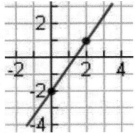

20.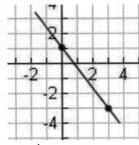

21.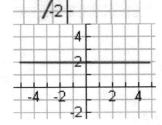

22.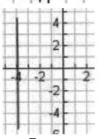

23. $y = 4$

24. $y = 6$

25. $x = 7$

26. $x = -3$

Chapter 7 : System of Two Linear Equations

7A – System of Two Linear Equations

The following is an example of a system of two linear equations in two unknowns (x and y):

$$x + y = 10$$

$$x - y = 4$$

The solution of the above system of equations is a pair, (x, y), whose coordinates sum to 10 and at the same time the $x - coordinate$ is 4 more than the y. The only such pair is (7,3):

$$7 + 3 = 10$$

$$7 - 3 = 4$$

In general we have a method for solving a system of two linear equations called the **Method of Elimination** that we outline below and demonstrate with a series of examples.

The Method of Elimination for Solving a System of Two Linear Equations in Two Unknowns:

1. Use basic operations (such as adding/subtracting the same expression to both sides or multiplying/dividing both sides by the same non-zero expression) to achieve:
   - both equations in the form $ax + by = c$.
   - Two $x - coefficients$ that are opposites (equal in absolute value with opposite signs), or two $y - coefficients$ that are opposites.

2. Add the two equations together to eliminate one of the variables; solve for the remaining variable.

EXAMPLE 1: Solve the system of equations: $x + y = 10$ and $x - y = 4$ by the method of elimination.

SOLUTION: Both equations are in the desired form, $ax + by = c$. The coefficients of y are opposites (1 and -1). Now add the two equations to eliminate y:

$$\begin{aligned} x + y &= 10 \\ \underline{x - y} &= \underline{\;\;4\;} \\ 2x\;\;\;\;\; &= 14 \\ x &= 7 \end{aligned}$$

To obtain the $y - coordinate$ of the solution, substitute $x = 7$ into either equation; we will substitute into $x + y = 10$.

$$\begin{aligned} 7 + y &= 10 \\ \underline{-7\;\;\;\;} &= \underline{-7} \\ y &= 3 \end{aligned}$$

The solution is $x = 7$ and $y = 3$, or in point form, (7,3) .

■

Note: The graph of each linear equation is a line. The solution of a system of linear equations is the intersection point of the two lines (assuming the lines are not parallel).

EXAMPLE 2: Find the x −value of the solution of the system of equations: $3x + 2y = 5$ and $5x − 4y = 3$.

SOLUTION: Since we want to find the x −value, we eliminate y.

$$2(3x + 2y) = (5)2 \qquad \leftarrow \text{Multiply both sides by 2 to achieve } y − coordinates \text{ that are}$$
$$5x − 4y \ = 3 \qquad\qquad \text{opposites (4 and } −4).$$

$$\begin{array}{r} 6x + 4y = 10 \\ \underline{5x − 4y = 3} \\ 11x \quad\ \ = 13 \end{array} \qquad \leftarrow \text{Add the equations together}$$

$$x = \frac{13}{11} \qquad \leftarrow \text{The } x −\text{value of the solution.} \quad \blacksquare$$

EXAMPLE 3: Find the x and y − values of the solution of the system: $3x = 5y − 9$ and $2x − 7y = −17$.

SOLUTION: First put the first equation in the form $ax + by = c$: $3x − 5y = 9$

$$(−2)(3x − 5y) = (−9)(−2) \qquad \leftarrow \text{Eliminate } x \text{ : multiply the top equation by } −2 \text{ and}$$
$$(3)(2x − 7y) = −17(3) \qquad\qquad \text{the bottom equation by 3 to achieve } x −$$
$$\qquad\qquad\qquad\qquad\qquad\qquad coordinates \text{ that are opposites.}$$

$$\begin{array}{r} −6x + 10y = 18 \\ \underline{6x − 21y = −51} \\ − 11y = −33 \end{array} \qquad \leftarrow \text{Add the equations together.}$$

$$y = 3 \qquad \leftarrow \text{Substitute } y = 3 \text{ into either equation to find } x.$$

$$3x = 5(3) − 9 \qquad \leftarrow \text{We substitute into the first equation.}$$
$$3x = 6$$
$$x = 2 \qquad\qquad\qquad\qquad\qquad\qquad \blacksquare$$

7A – EXERCISES

For $1 − 8$, solve for the indicated variable.
1. $2x + y = 3, 2x − y = 7$; solve for x (eliminate y).
2. $4x − 3y = 2, −4x + 5y = 4$; solve for y (eliminate x).
3. $3x + 2y = 4, 5x − y = 3$; solve for x.
4. $6x − 2y = 5, 3x + 5y = 7$; solve for y.
5. $5x − 3y = 7, 15x + 2y = 3$; solve for y.
6. $3x = 4y + 2, 5x − 2y = 1$; solve for x (put the first equation in the form $ax + by = c$).
7. $2x + 5y = 1, −3x + 7y = 3$; solve for y.
8. $5x − 3y = 2, 4x + 5y = −4$; solve for x.

For $9 − 10$, solve for both x and y.
9. $−3x + 2y = −7, x − 5y = 11$
10. $5x − 3y = −11, 7x + 4y = 1$

7A – WORKSHEET: System of Two Linear Equations

For 1 – 8, solve for the indicated variable.

1.	$5x + 2y = 3, 3x - 2y = 4$; solve for x (eliminate y).
2.	$5x - 4y = 1, -5x + 2y = 3$; solve for y (eliminate x).
3.	$2x + 3y = -1, 2x - y = 5$; solve for y.
4.	$4x - 3y = 1, 2x + 5y = -6$; solve for y.
5.	$3x - 7y = 5, 9x + 2y = 4$; solve for y.
6.	$2x = 3y + 4, 5x - 6y = -1$; solve for x (put the first equation in the form $ax + by = c$).

7. $5x + 4y = 1, -3x + 2y = 1$; solve for y.

8. $7x - 2y = 2, 4x + 3y = -2$; solve for x.

For $9 - 10$, solve for both x and y.

9. $3x - 4y = -10, 6x + 5y = -7$

10. $2x - 3y = -13, 4x + 5y = 7$

Answers:

1. $x = \dfrac{7}{8}$

2. $y = -2$

3. $y = -\dfrac{3}{2}$

4. $y = -1$

5. $y = -\dfrac{11}{23}$

6. $x = -9$

7. $x = -\dfrac{1}{11}$

8. $x = \dfrac{2}{29}$

9. $x = -2, y = 1$

10. $x = -2, y = 3$

7 – Answers to Exercises

Section A

1. $x = \dfrac{5}{2}$

2. $y = 3$

3. $x = \dfrac{10}{13}$

4. $y = \dfrac{3}{4}$

5. $y = -\dfrac{18}{11}$

6. $x = 0$

7. $y = \dfrac{9}{29}$

8. $x = -\dfrac{2}{37}$

9. $x = 1, y = -2$

10. $x = -1, y = 2$

Practice Final Exam - One

1.	$7\sqrt{24} - 5\sqrt{6}$ a) $9\sqrt{6}$ b) $12\sqrt{2}$ c) $\sqrt{24}$ d) $4\sqrt{3}$
2.	$\sqrt{2}(\sqrt{2} + \sqrt{8})$ a) 6 b) $2 + \sqrt{10}$ c) $3\sqrt{2}$ d) $6 + 2\sqrt{2}$
3.	$\dfrac{\sqrt{7}\sqrt{35}}{\sqrt{5}}$ a) 7 b) $\sqrt{7}$ c) $\sqrt{5}$ d) $\sqrt{35}$
4.	$\dfrac{-21a^6 b^5}{-3a^2 b^2}$ a) $-7a^3 b^4$ b) $7a^4 b^3$ c) $-7a^4 b^2$ d) $7a^3 b^2$
5.	$(2x^2 y^3)^3$ a) $6x^5 y^6$ b) $8x^6 y^9$ c) $8x^6 y^6$ d) $6x^8 y^9$
6.	$(5x^2 - 7x + 6) - (-2x^2 + 6x - 5)$ a) $7x^2 - 13x + 11$ b) $3x^2 - 13x + 11$ c) $7x^2 + 13x - 11$ d) $7x^2 - 13x - 11$
7.	$(2x - 3)(x^2 + 3x - 5)$ a) $2x^3 - 3x^2 + 21x - 15$ b) $2x^3 + 3x^2 - 21x - 15$ c) $2x^3 + 3x^2 + 19x + 15$ d) $2x^3 + 3x^2 - 19x + 15$
8.	$\dfrac{4x^3 - 6x^2 + 2x}{2x}$ a) $2x^2 + 3x + 1$ b) $2x^3 - 6x + 1$ c) $2x^2 - 3x + 1$ d) $2x^2 - 3x - 1$
9.	Factor completely: $27x^3 - 12xy^2$ a) $3xy(9x - 4)$ b) $3x(9x + 2y)(9x - 2y)$ c) $3xy(2x^2 - 4y^2)$ d) $3x(3x + 2y)(3x - 2y)$
10.	Which of the following is a factor of the polynomial $2x^2 + 11x - 21$ a)$x - 7$ b)$x + 7$ c)$2x - 7$ d) $2x + 3$
11.	Which of the following is a factor of the polynomial? $6ac - 4ad + 9bc - 6bd$ a) $2a + 3b$ b) $2a - 3b$ c) $3c + 2d$ d) $6a + 2d$
12.	If y represents a number , which equation is a correct translation of the sentence: **12 is 3 less than twice a number.** a) $2y - 3 = 12$ b) $2y - 12 = 3$ c) $y - 3 = 12$ d) $3 = 12 - y$
13.	Solve for x: $\dfrac{x-3}{2} + \dfrac{1}{6} = \dfrac{7}{6}$. a) $x = -3$ b) $x = 3$ c) $x = 4$ d) $x = 5$
14.	Solve for n: $18 - 6n = 3(2 - n)$ a) $n = -4$ b) $n = 4$ c) $n = 6$ d) $n = -6$
15.	Which is the value of the $y - coordinate$ of the solution to the system of equations? $2x + 3y = 13, x + y = 5$ a) $y = -3$ b) $y = 2$ c) $y = 3$ d) $y = -2$
16.	Solve for x: $z = 7x - 14y$ a) $x = \dfrac{z - 14y}{7}$ b) $x = \dfrac{z + 7y}{14}$ c) $x = \dfrac{z + 14y}{7}$ d) $x = \dfrac{7z + 14y}{7}$
17.	Find all the solutions of the equation: $x^2 + 2x = 8$ a) $x = 4, x = 3$ b) $x = 4, x = -2$ c) $x = -4, x = 2$ d) $x = 6, x = 2$

(practice for final exam – one)

18.

Find the value of x in the right triangle.

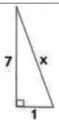

a) $5\sqrt{3}$ b) $3\sqrt{2}$ c) $2\sqrt{3}$ d) $5\sqrt{2}$

19. Find the graph of the solution to the inequality: $2x - 3 < 5x - 6$

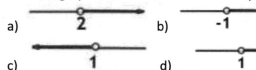

a)   b)

c)   d)

20. Given $a = 4$ and $b = -5$, evaluate the expression given below:
$$3a^3 - 5b^2 - 2ab$$
a) 110 b) 117   c) 120   d) 107

21. Which of the following is the graph of the equation : $2x + 3y = 12$

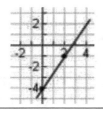

a)   b)   c)   d)

22. Find the equation of the line through points $(-1,4)$ and $(2,-2)$.
a) $y = 2x - 2$ b) $y = -2x - 2$ c) $y = 3x + 2$ d) $y = -2x + 2$

23. Find the equation of the vertical line passing through point $(-5,4)$.
a) $x = 4$ b) $x = -5$ c) $y = 4$ d) $y = -5$

24. Find the slope and the $y - intercept$ for the graph of equation: $4x - 6y = 42$
a) slope $= \frac{3}{2}$ y-intercept $= -7$ b) slope $= \frac{2}{3}$ y-intercept $=$ 7
c) slope $= \frac{2}{3}$ y-intercept $= -7$ d) slope $= 2$ y-intercept $=$ -7

25.

What is the slope of the line?
a) $-\frac{4}{3}$ b) $\frac{4}{3}$ c) 7   d) $-\frac{3}{4}$

1.	a	6.	a	11.	a	16.	c	21.	c
2.	a	7.	d	12.	a	17.	c	22.	a
3.	a	8.	c	13.	d	18.	d	23.	b
4.	b	9.	d	14.	b	19.	d	24.	c
5.	b	10.	b	15.	c	20.	d	25.	a

Practice Final Exam - Two

1.	$6\sqrt{48} - 3\sqrt{12}$ a) $3\sqrt{36}$ b) $18\sqrt{3}$ c) $16\sqrt{2}$ d)$16\sqrt{3}$
2.	$\sqrt{3}(\sqrt{3} + \sqrt{27})$ a)6 b) 12 c) $3\sqrt{10}$ d) $3 + \sqrt{30}$
3.	$\frac{\sqrt{10}\sqrt{60}}{\sqrt{6}}$ a) $4\sqrt{6}$ b)$10\sqrt{6}$ c) $6\sqrt{10}$ d) 10
4.	$\frac{-72x^6y^4}{6x^2y^3}$ a) $-12x^4y$ b) $-12x^3y$ c)$12x^3y$ d) $12x^4y$
5.	Simplify: $(3x^2y^3)^4$ a)$27x^2y^{12}$ b) $81x^8y^{12}$ c) $27x^8y^7$ d) $12x^8y^{12}$
6.	$(3x^2 + 5x - 2) - (-4x^2 - 2x + 7)$ a) $7x^2 + 7x - 9$ b) $-7x^2 + 7x - 9$ c) $7x^2 - 7x + 9$ d)$7x^2 + 7x + 9$
7.	Multiply: $(3x - 1)(3x^2 - 2x - 5)$ a) $9x^3 + 9x^2 + 15x + 15$ b) $9x^3 + 9x^2 - 13x + 5$ c) $9x^3 - 9x^2 + 13x + 5$ d) $9x^3 - 9x^2 - 13x + 5$
8.	Simplify: $\frac{72x^3 - 24x^2 + 12x}{-12x}$ a) $-6x^2 - 2x + 1$ b) $-6x^2 + 2x - 1$ c) $-6x^2 - 2x - 1$ d)$-6x^2 + 2x + 1$
9.	Factor completely: $98x^3y^3 - 50xy$ a)$2xy(2xy + 25)(2xy - 25)$ b) $2xy(49x^2y^2 - 5)$ c) $2xy(7xy + 5)(7xy - 5)$ d) $2x(7x + 25y)(7x - 25y)$
10.	Which of the following is a factor of $2x^2 - x - 6$ a) $x + 6$ b) $x - 2$ c) $2x - 3$ d) $x + 2$
11.	Which of the following is a factor of the polynomial $6ac - 4ad + 3cb - 2bd$? a) $2a - b$ b)$a + 2b$ c) $2a + b$ d)$3c + 2d$
12.	If y represents a number, which equation is a correct translation of the following sentence? **12 subtracted from 10 times a number is 5.** a) $12 - 10y = 5$ b) $10y - 12 = 5$ c) $12y - 10 = 5$ d) $10 - 12y = 5$
13.	Solve for x: $\frac{x-4}{2} = \frac{x-7}{3}$ a)$x = 2$ b) $x = 3$ c) $x = -2$ d)$x = -1$
14.	Solve for n: $3(6 - 2n) = 6$ a) $n = 2$ b) $n = -2$ c) $n = 3$ d) $n = -3$
15.	What is the x-coordinate of the solution of the system of equations: $2x + y = 10, -x + 2y = 7$ a) $x = \frac{13}{5}$ b)$x = -\frac{3}{5}$ c) $x = 3$ d) $x = -13$
16.	Solve for x: $z = 3x + 2y$ a) $x = \frac{z+2y}{3}$ b) $x = \frac{z-2y}{3}$ c) $x = \frac{3z-y}{3}$ d) $x = \frac{z+2y}{6}$
17.	Find all the solutions to the equation: $x^2 + x = 2$ a)$x = -1, x = 2$ b) $x = 3, x = -1$ c) $x = 4, x = 1$ d) $x = 1, x = -2$
18.	Given $x = 5$ and $y = -2$, evaluate the expression given below: $3x^2 - xy$. a) 85 b) 75 c)70 d) 62

(practice for final exam – two)

19.

What is the value of x in the right triangle:

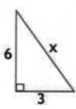

a) $6\sqrt{3}$ b) $3\sqrt{5}$ c) $5\sqrt{3}$ d) $2\sqrt{3}$

20. Find the graph of the solution of the inequality: $4x + 6 \geq 7x - 3$.

a) **3**   b) **-1**   c) **3**   d) **-1**

21. Which of the following is the graph of the equation $2x - y = 6$.

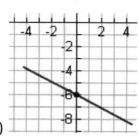 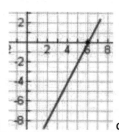 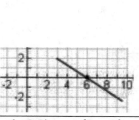 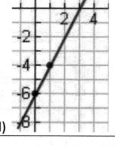

a)   b)   c)   d)

22. Find the equation of the line passing through points $(-2,8)$ and $(3,-7)$.
a) $y = 3x - 2$ b) $y = 5x + 3$ c) $y = -3x + 2$ d) $y = -2x + 8$

23. Find the equation of the horizontal line passing though the point $(-5,4)$.
a) $x = -5$ b) $x = 4$ c) $y = x + 4$ d) $y = 4$

24. Find the slope and y-intercept of the line with equation: $5x + 3y = 15$
a) $m = 5; (0,15)$ b) $m = -\frac{5}{3}; (0,-5)$ c) $m = -\frac{5}{3}; (0,5)$ d) $m = \frac{5}{3}; (0,-2)$

25.

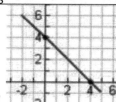

What is the slope of the line ?
a) 1 b) -1 c) 4 d) -2

1.	b	6.	a	11.	c	16.	b	21.	d
2.	b	7.	d	12.	b	17.	d	22.	c
3.	d	8.	b	13.	c	18.	a	23.	d
4.	a	9.	c	14.	a	19.	b	24.	c
5.	b	10.	b	15.	a	20.	c	25.	b

Practice Final Exam – Three

1.	$3\sqrt{75} - 4\sqrt{12}$ a) $7\sqrt{3}$ b) $-\sqrt{2}$ c) $15\sqrt{3}$ d)$15\sqrt{3} - 8\sqrt{2}$
2.	$\sqrt{5}(\sqrt{5} + \sqrt{20})$ a)$5\sqrt{5}$ b) $3\sqrt{5}$ c) 5 d) 15
3.	$\frac{\sqrt{3}\sqrt{30}}{\sqrt{10}}$ a) $3\sqrt{10}$ b) 9 c) $2\sqrt{30}$ d) 3
4.	$\frac{-14a^8b^5}{-7a^4b}$ a) $2a^4b^4$ b) $-2a^4b^4$ c) $-2a^2b^5$ d) $2a^2b^5$
5.	Simplify: $(5x^2y^3)^3$ a) $125x^6y^9$ b) $15x^5y^6$ c) $125x^5y^6$ d) $25x^6y^9$
6.	$(6x^3 - 7x^2 + 6x + 2) - (-3x^3 - 2x^2 + 7x + 9)$ a) $3x^3 - 5x^2 + x - 7$ b) $9x^3 + 5x^2 + x + 7$ c) $9x^3 - 5x^2 - x - 7$ d) $9x^3 - 5x^2 - x + 7$
7.	Multiply: $(2a + 5)(-3a^2 + a - 2)$ a) $6a^3 - 13a^2 + a + 10b$ b) $-6a^3 + 13a^2 - a + 10$ c) $-6a^3 - 13a^2 + a - 10$ d) $-6a^3 - 13a^2 - a + 10$
8.	Simplify: $\frac{12x^3+6x^2-9x}{-3x}$ a) $-4x^2 - 2x + 3$ b) $-4x^2 + 2x - 3$ c) $-4x^2 + 2x + 3$ d)$4x^2 + 2x + 3$
9.	Factor completely: $18a^3b - 50ab$ a) $2ab(9a^2 - 5)$ b) $2ab(3a + 5)(3a + 5)$ c) $2ab(3a + 5)(3a - 5)$ d) $2a(9b - 25)$
10.	Which of the following is a factor of the polynomial: $2x^2 + 9x - 18$ a)$x + 6$ b) $2x + 3$ c) $x + 3$ d) $x - 6$
11.	Which of the following is a factor of the polynomial: $15xw - 10xz + 6yw - 4yz$ a)$5x - 2y$ b) $2x - 5y$ c) $5x + 2y$ d) $3w + 2z$
12.	If y represents a number which equation is a correct translation of this sentence: **15 added to twice a number is 25** a) $2y + 15 = 25$ b)$2y - 15 = 25$ c) $2y + 25 = 15$ d) $y + 15 = 2(25)$
13.	Solve for y: $\frac{y-3}{4} + \frac{4}{3} = \frac{7}{3}$ a) $y = 1$ b) $y = -1$ c) $y = -7$ d) $y = 7$
14.	Solve for n: $5(7 - n) = 4n - 25$ a) $n = \frac{20}{3}$ b) $n = 4$ c) $n = \frac{21}{4}$ d) $n = -7$
15.	What is the value of the x-coordinate of the solutions to the system of equations: $2x - 3y = 8, 3x + y = 1$ a) $x = -1$ b) $x = 1$ c) $x = 2$ d) $x = 3$
16.	Solve for x: $z = 5x - 10y$ a) $\frac{z-10y}{5}$ b) $\frac{10y-5x}{2}$ c) $\frac{z+10y}{5}$ d) $\frac{z+10y}{10}$
17.	Find all solutions to the equation: $x^2 + 2x = 24$ a) $-6, 4$ b) $4, 6$ c) $-6, -4$ d) $6, -4$
18.	Given $x = 5$ and $y = -6$ evaluate the expression given below: $3x^2 + 2xy + 2y^2$ a)75 b) - 60 c) 87 d) 15
19.	Find the equation of the line passing through points $(0,3)$ and $(-1,4)$. a) $y = x + 3$ b) $y = -x - 3$ c) $y = 3x - 1$ d) $y = -x + 3$

	(practice final exam – three)
20.	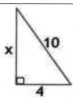 What is the value of x in the right triangle ? a) $2\sqrt{21}$ b) $21\sqrt{2}$ c) $10\sqrt{3}$ d)$4\sqrt{10}$
21.	Find the graph of the solution to the inequality: $2x + 3 \geq 6x - 5$ 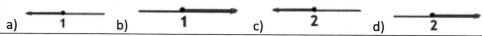 a) 1 b) 1 c) 2 d) 2
22.	Which of the following is the graph of $5x + 2y = -10$ 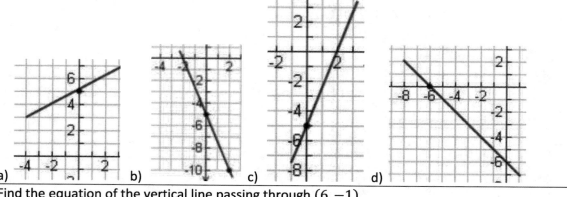 a) b) c) d)
23.	Find the equation of the vertical line passing through $(6, -1)$. a) $x = -1$ b) $y = 6$ c) $y = -1$ d) $x = 6$
24.	Find the slope and y-intercept: $6x + 4y = -12$ a) $m = -\frac{3}{2}; (0, -3)$ b) $m = -\frac{3}{2}; \left(0, \frac{1}{3}\right)$ c) $m = \frac{3}{2}; (0,3)$ d) $m = -\frac{3}{2}; \left(0, -\frac{1}{3}\right)$
25.	What is the slope of the line? a) $-\frac{3}{2}$ b) $-\frac{2}{3}$ c) $\frac{2}{3}$ d) $\frac{3}{2}$

1.	a	6.	c	11.	c	16.	c	21.	c
2.	d	7.	c	12.	a	17.	a	22.	b
3.	d	8.	a	13.	d	18.	c	23.	d
4.	a	9.	c	14.	a	19.	d	24.	a
5.	a	10.	a	15.	b	20.	a	25.	d